THE KEY ISSUES LECTURE SERIES
is made possible through a grant from
International Telephone and Telegraph Corporation

Labor, Technology, and Productivity in the Seventies

Edited by
Jules Backman

With a Foreword by
Harold S. Geneen

New York: New York University Press   1974

Library of Congress Catalog Card No.: 74-13004
ISBN: cloth: 0-8147-0981-8
ISBN: paper: 0-8147-0982-6

Manufactured in the United States of America

Preface

Abraham L. Gitlow
Dean
College of Business and Public Administration
New York University

The appearance of the third volume of the Key Lecture Series of the College of Business and Public Administration, New York University, is a source of genuine pleasure and no little pride. Of course we are grateful to the International Telephone and Telegraph Corporation for the grant which made this Lecture Series possible. This latest series of lectures focused on issues involving the trade union movement in the United States, its interrelationships with business, the consequences of these interrelationships for industrial peace or strife, and the prospects for technological progress and its momentum so that the productivity of our economy would be advanced in the future. Once again, some of the best and most knowledgeable minds in the country were invited to express their thoughts on the Key Issues mentioned above. It is a certainty that the opinions expressed by Professors Emanuel Stein, Peter F. Drucker, and John W. Kendrick as well, as the nonprofessorial but extraordinarily experienced opinion of Mr. David L. Cole, will be read and discussed widely. In fact, this has already been the case.

I would be remiss indeed if I did not express appreciation to Professor Jules Backman who was instrumental in arranging the Series; to his Secretary, Mrs. Catherine Ferfoglia, who helped to prepare the volume for publication; to my Administrative Assistants, Mrs. Patricia Matthias and Mrs. Virginia Moress; and to Mr. Malcolm Johnson and Mr. Robert Bull of the NYU Press, who followed the manuscript to timely publication.

5

Contents

Foreword

Harold S. Geneen

Chairman and Chief Executive
International Telephone and Telegraph Corporation

The third Key Issues Lecture Series in the spring, of 1974, deals with our most valuable resource—the approximately 90 million Americans who make up our labor force. This human factor must underlie any consideration of the tremendous technological, economic, and psychological changes taking place in today's marketplace. While the world shrinks in size, the job of coping with these changes grows larger and more important with each passing day.

In fact, with all of the questions and problems and uncertainties that span today's globe, you might say that we actually face a *world* in transition—as attitudes and life styles change as never before.

Doomsayers take that to mean that Modern Man has exploited his resources and exceeded his capabilities to such a degree that decline or self-destruction is inevitable. These harbingers of gloom point to worldwide energy shortages and to economic dislocation brought about by so-called runaway technology.

Those of us at ITT however, see another—much brighter—side to the global coin.

While it is undeniably true that we face tougher, wider ranging problems than ever before, *it is also true* that these problems offer each and every one of us—labor, management, consumer, student, educator and politician alike—the *greatest opportunity ever* to put our educational and scientific tools to work to prove that unity and resolution can surmount any obstacle.

The first step is recognizing and admitting the seriousness of the problem we face. That is what this current group of lectures presented here as *Labor, Technology, and Productivity in the Seventies* tries to do.

Once again we have had the outstanding leadership of Dr. Jules Backman, Research Professor of Economics, who directs the NYU series. I am most gratified that our lectures are contributing to the public dialogue in a very definite way.

Tables

Tables

Labor, Technology, and Productivity in the Seventies

ONE

Emerging Trends

Jules Backman

Research Professor of Economics
New York University

Technology, productivity, and labor are closely interrelated. Advances in technology contribute significantly to productivity, whether measured in terms of man-hour input (OPM) or factor productivity (capital plus labor input).[1] The rates of increase in OPM influence significantly the level of real wages and the extent to which higher hourly labor costs will affect unit labor costs (ULC). When hourly labor costs rise faster than OPM, the resulting higher unit labor costs are described as wage inflation or labor cost inflation or cost-push inflation. Depending upon the trends in the economy, cost-push inflation may lead to higher prices, a profit squeeze, and/or unemployment.[2]

During the post-World War II period, cost-push periodically has been a significant problem (1946-48, 1956-58, 1966-74). Table 1-1 shows the changes in ULC for the private nonfarm economy.

With only four exceptions, unit labor costs have increased in every year since 1946. Wage costs have outstripped the large increases in OPM despite the great strides in technology and the enormous increase in expenditures for research and development. (See Table 1-6.) One result has been a significant rise in labor's share of national income. (See Table 1-4.)

17

Table 1-1

COMPENSATION PER MAN-HOUR, OUTPUT PER MAN-HOUR AND
UNIT LABOR COSTS IN THE PRIVATE NON-FARM ECONOMY, 1947–1973

	Compensation per Man-hour	Output per Man-hour (1967 = 100)	Unit Labor Costs	Year-to-Year Changes in Unit Labor Costs
1947	38.3	57.1	67.1	
1948	41.8	58.8	71.0	+ 5.8
1949	43.0	61.1	70.3	−1.0
1950	45.3	65.0	69.7	−0.8
1951	49.3	66.3	74.3	+ 6.6
1952	52.0	66.9	77.6	+ 4.5
1953	54.9	68.9	79.7	+ 2.6
1954	56.6	70.5	80.3	+ 0.9
1955	58.6	73.6	79.6	−0.9
1956	62.0	73.2	84.7	+ 6.4
1957	65.5	74.8	87.6	+ 3.4
1958	68.1	76.7	88.7	+ 1.3
1959	71.0	79.3	89.5	+ 0.9
1960	73.9	80.3	92.0	+ 2.8
1961	76.3	82.7	92.3	+ 0.2
1962	79.3	86.4	91.8	−0.5
1963	82.2	89.1	92.3	+ 0.5
1964	86.1	92.4	93.2	+ 1.0
1965	89.2	95.1	93.9	+ 0.8
1966	94.6	98.4	96.2	+ 2.5
1967	100.0	100.0	100.0	+ 4.0
1968	107.3	102.9	104.3	+ 4.3
1969	114.8	102.7	111.8	+ 7.2
1970	123.2	103.4	119.1	+ 6.6
1971	131.8	107.6	122.5	+ 2.8
1972	140.9	112.1	125.1	+ 2.6
1973[p]	151.6	115.5	131.2	+ 4.4

Source: *Economic Report of the President*, February 1974 (Washington: 1974), pp. 286–87.

This is the background against which the papers included in this volume should be analyzed. The concern of the authors is with the emerging trends and how they will affect future developments.

LABOR IN TRANSITION

The short-term objectives of labor unions in this country have changed considerably over the past forty years. As each goal has been achieved, new areas have been staked out. The New Deal marked a significant turning point in the evolution of American labor and labor unions. Prior to 1933, organized labor was confined largely to a few industries (e.g., apparel and coal) and to crafts in other industries (e.g., printing, construction, and railroads). The predominant form of union organization was the craft union. About one worker in ten belonged to a union. The major thrust of American trade unions following Samuel Gompers's advice was to press on "bread-and-butter" issues such as higher wages, shorter hours, and improved working conditions.

In the 1930s, several major pieces of legislation [3] and the Supreme Court decisions upholding the Wagner Act strongly encouraged the labor movement.

Union Membership

The Wagner Act provided a major opportunity for organized labor, since it guaranteed the right to organize and to bargain collectively as part of our national policy. Thus, it is not surprising that in no other period in our history did the trade union movement grow as rapidly. A split in the labor movement led to the creation of the CIO and major organizing drives by the newly created industrial unions in steel, automobiles, rubber, and other mass-production industries. This was the main thrust during the 1930s and World War II.

The major increase in union membership during this period is shown in Table 1-2.

Table 1-2

NONAGRICULTURAL EMPLOYMENT AND UNION MEMBERSHIP,
SELECTED YEARS, 1930–1972

	Total No. of Wage & Salary Workers (millions)	Total Union Membership	Percent
1930	29.4	3.4	11.6
1939	30.6	8.8	28.8
1947	43.9	14.8	33.7
1957	52.9	17.4	32.9
1965	60.8	17.3	28.5
1970	70.6	19.4	27.5
1972	72.8	19.4	26.6

Source: U.S. Department of Labor, Bureau of Labor Statistics.

By 1947, one worker in three belonged to a labor union. There was no basic change in this ratio in the following ten years. Although total union membership increased in the 1960s, the total represented a declining proportion of the rapidly expanding labor force. Currently, about one worker out of four is a member of a labor union.

Influence of Labor Unions

These relationships understate the role and influence of labor unions. First, a union negotiates for an entire bargaining unit even though not all workers are members of the union. As Professor Stein states, "The influence of the unions is in many instances, therefore, considerably wider than the size of the union membership would suggest." Secondly, the wage and nonwage benefits negotiated by major unions become the "key bargains," which are followed after varying periods of delay by unorganized workers. Third, although organized labor always has represented a minority of the workers, it has been responsible for most of the new initiatives in the area of nonwage benefits and governmental action.

During World War II, the unions successfully insisted upon participating in the operation of various aspects of the war effort and received warm support from President Roosevelt. The concept of a tripartite approach (labor, management, and the public) was adopted by the National War Labor Board.

In the early postwar period, the major thrust was directed to so-called fringe benefits including vacations, holidays, pensions and welfare funds. In addition, greater emphasis was given to political activity. However, labor unions in this country have never achieved the political status of labor in England and in some other countries. At the end of the first postwar decade, the AFL and CIO merged.

In 1959, the Labor-Management Reporting and Disclosure (Landrum-Griffin) Act was passed. Of particular importance were its restrictions on the power of union leaders and the concomitant increase in union democracy for the membership. (Titles I, III, and IV). The results have included an increasing number of grievances and a number of instances in which a militant membership has refused to ratify agreements reached by its leaders. Professor Stein underlines the "marked and widespread rank-and-file restiveness" and the "unusual intransigence on the part of the membership." This has added a new dimension of uncertainty to the collective-bargaining process because it has placed labor leaders under enormous pressure to seek to satisfy all demands of the membership—no matter how unwarranted—if the leaders are to hold on to their jobs. "Militancy of the membership translates itself quickly into militancy of the leadership."

During the past decade, the main thrust of labor unions has been upon organizing the public sector,[4] upon getting wage increases to compensate for price inflation, upon liberalizing existing nonwage benefits, upon political activity to assure labor's participation in public policy, and upon legislation to improve the economic position of workers (e.g., higher minimum wages, improved social security benefits, occupational health and safety laws, national health programs, larger unemployment benefits for longer periods).

Labor Gains

The past four decades have witnessed major gains in levels of living for workers, due primarily to the large expansion in national economic activity and to a small extent to pressure by labor unions. The increase in real weekly wages, adjusted for the price inflation, is shown in Table 1-3.

Table 1-3

PRIVATE NONAGRICULTURAL REAL WEEKLY EARNINGS
(in 1967 dollars)

	Real Gross Weekly Earnings	Real Spendable Weekly Earnings [1]
1947	$ 68.13	$66.73
1957	86.99	80.32
1965	100.59	91.32
1969	104.38	91.07
1970	102.72	89.95
1971	104.62	92.43
1972	108.36	96.40
1973	108.43	95.08
Annual Rate of Increase:		
1947–65	2.2%	1.8%
1965–73	0.9	0.4
1947–73	1.8	1.4

1. Workers with three dependents, after federal income taxes.

The major gains in real weekly earnings were recorded largely before the new price inflation started in 1965. Between 1947 and 1965, real weekly earnings before taxes increased at an annual rate of 2.2 percent and between 1965 and 1973 at an annual rate of only 0.9 percent. On an after-tax basis, the disparity is even wider (1.8 percent and 0.4 percent), largely because of the sharp rise in social security taxes in recent years. There has been a marked slowdown in labor's gains since 1965. However, in addition to the gains in real weekly earnings, the average length of the work week declined by 1.5 hours between 1947 and 1965 and by 1.7 hours in the following eight years. Moreover, during the 1947-73 period, enormous improvements also were obtained in various nonwage benefits, so that the real improvement in the economic position of workers is greater than indicated by the changes in real weekly earnings alone.[5]

Labor's share of national income has increased significantly during the post-World War II period. (See Table 1-4.)

Table 1-4

LABOR'S SHARE OF NATIONAL INCOME

	Total National Income (billions)	Compensation of Employees	% of Total
1947	$ 199.0	$128.9	64.8
1957	366.1	256.0	69.8
1965	564.3	393.8	72.4
1968	711.1	514.6	74.5
1973	1054.2	785.3	74.5

Source: *Economic Report of the President,* February 1974 (Washington: Council of Economic Advisers, 1974), p. 266.

Diverse Objectives of Unions

Labor is not a monolith with unified objectives. As Professor Stein notes, labor is a "very centralized affair"—a collection of unions each having autonomy and diverse objectives. There are unions with maturity and a long history of collective bargaining (the railroad unions) and unions which are relatively new in areas where they have been catching up with the benefits already obtained by the older unions (hospital workers); unions which represent all of the workers in an industry (auto workers) and unions which represent only a small group of workers in a local company (construction union locals); unions which already have organized an entire industry (steelworkers) and unions seeking to become established (United Federation of Teachers in higher education).

There are, also, unions concerned about import competition and the effect on job opportunities (electrical workers) and those which may benefit from imports (longshoremen); unions in sectors where strikes may not be tolerable (police and firemen) and unions whose strikes may merely be a minor inconvenience (retail clerks); unions which are innovative in their demands (autoworkers) and unions which just seek to obtain the gains realized by the innovators (most small unions); unions in industries where labor costs account for a very large proportion of the sales dollar (coal miners) and those in industries where such costs are a small percentage (petroleum workers); unions which are militant politically (municipal workers in New York City) and unions which usually play a more modest role in politics (most unions); unions in dynamic, expanding industries (chemical workers) and those in industries which have been declining in relative importance (textile workers).

Moreover, since labor unions represent only slightly more than one-fourth of all wage and salary workers, they cannot speak with a single voice for all workers even when the unions have an objective on which most can agree (e.g., the opposition to the Taft-Hartley Act). Against this background, it is clear that emerging and future trends in the labor movement will not be of uniform interest to all workers and will have diverse impacts on and responses from the organized sectors.

Nevertheless, when broad new trends develop, they sweep across the entire spectrum of labor—sometimes slowly but inexorably (nonwage benefits). And all workers feel the impact of price inflation and many would be affected by mass unemployment, as we saw during the 1930s. Such developments affect the bread-and-butter issues of jobs and purchasing power and hence are of interest to all workers—organized and unorganized. But that even in these areas the approaches used are not uniform is clearly shown by the fact that only a little more than 4 million workers are covered by automatic cost-of-living adjustment clauses; others depend upon more frequent reopening of contracts or seek to catch up for losses in purchasing power in the future negotiations.

The power structure of the unions, their diverse objectives, the extent to which they can unify members behind a goal, their maturity, the scope of their representation, and a number of other factors will affect the timing and extent to which future developments will affect each union.

Future Trends

Where does organized labor go from here? The dynamics of the labor movement require the achievement of new gains—or the leadership will be replaced. Where can these gains be sought?

Under the pressure of price inflation higher wages will continue to be the prime objective of workers. However, as Professor Stein concludes, this will be accompanied "by the drive for expanded fringe benefits." The bread-and-butter issues thus continue to provide the major thrust in collective bargaining.

Several portents for the future are indicated by the developments in the early 1970s. Labor unions traditionally have opposed arbitration of new contracts although they have generally accepted arbitration for grievances. The no-strike agreement between the United Steelworkers and the steel industry in March 1973, more than a year before its contract was to be reopened, to submit to binding arbitration unresolved issues,[6] may represent a breakthrough in this area. In April 1974, an agreement was reached on a new contract without recourse to arbitration. The parties agreed to follow the same procedure when the contract expires in 1977.

It doesn't make much sense to engage in long strikes, with great inconvenience and hardship to the workers, the industry, and the public, over the last few cents of a wage increase or over other issues which can be settled by third-party intervention.[7] Moreover, strikes by hospital employees, police, firemen, etc., affect vital services and cannot be tolerated. David Cole notes a number of actions which may be taken including contractual agreements to arbitrate. He is opposed to legislatively mandated arbitration.

The continuing inflation since 1965 has given greater impetus to the adoption of cost-of-living clauses.[8] In 1972, Congress provided for automatic annual adjustments in social security benefits based on changes in the Consumer Price Index (CPI).[9] In 1974, cost-of-living adjustments for pensions were agreed upon in the steel, aluminum, and can industries. Such provisions will spread to other private pension programs like wildfire. However, if these private pension funds are to remain solvent, the cost-of-living provision could require a substantial allocation of funds for this purpose, thus reducing to some extent the amount available to pay wage increases or to improve other nonwage benefits for those who still are working.

New initiatives dealing with noneconomic elements of the labor contract are indicated by the United Automobile Workers' emphasis upon reducing the speed of the assembly line and limiting the extent of overtime that may be demanded by management. The improvements in health and safety in mining presage greater emphasis upon this area in other industries. Traffic congestion may lead to some efforts to stagger hours.

One area that has not yet developed but may be in the cards in the years ahead is a new drive for a reduced work week. Some companies have experimented with the four-day week, and the length of the work week has been reduced to thirty-five hours in some industries. If the annual rate of economic growth slows up in the years ahead while the expansion in the work force continues, as seems probable, there is bound to be increasing demand to spread the available work by reducing the length of the work week as one way to avoid increasing unemployment.

The changing nature of our economy, with service industries accounting for a steadily larger share of total output, undoubtedly will affect union organizing activity and may contribute to a weakening of the relative strength of the unions which are so largely based in manufacturing and mining—although unionization of government employees has increased significantly. The increasing number of women in the labor force creates opportunities as well as problems for trade unions. Women have not flocked into labor organizations in the past. The challenge is clear for unions, but the past record is poor.

IS INDUSTRIAL PEACE ACHIEVABLE?

Tens of thousands of labor contracts are negotiated peaceably each year. Except for contracts with major companies in steel, autos, railroads, and a few other key industries and in some local industries, such peaceful settlements are not news and hence usually are not called to public attention. And even many negotiations which are characterized by a considerable amount of public bickering are settled without resort to a strike.

Moreover, when the parties cannot agree in key sectors, government may step in and through mediation or public fact-finding boards seek to create a new atmosphere in which an agreement can be reached. Provisions of the Taft-Hartley and Railway Labor Acts make possible the postponement of a threatened shutdown for a period of time and often lead to an agreement without a strike.

Is industrial peace achievable? In the overwhelming majority of negotiations each year, the answer is yes. However, even when an agreement is reached without a strike, there may be considerable turbulence in labor-management relations. As David Cole states, "The number of man days lost in a stipulated period does not necessarily reflect the state of labor relations."

Resentment may build up because of working conditions, methods of handling grievance procedures, wage differentials for skilled workers, excessive overtime, pending technological changes which result in job displacement, and so on. Thus, beneath the apparently peaceful surface, conditions may be warlike and lead to an eventual explosion. As the sore festers, there may be a negative impact on morale and there is the growing likelihood of worker alienation. The resulting "Friction or a sense of hostility at the workplace causes inefficiency."

Impact of Strikes

The overt evidence of a lack of industrial peace is found in the strike or the lockout. Management and labor cannot always reach agreement concerning the terms of a new contract or the handling of grievances. The result may be a strike either called by a union or concerted action by some workers without the approval of the union (wildcat strike).

Each year millions of man-days of work are lost because of strikes. (See Table I-5.) The number of days lost each year, however, usually is much less than 1 percent of the time worked, although it has been that high in an unusual year such as 1946. Generally, the proportion of man-days lost has been less than one-quarter of one percent.[10]

Strikes have varying impacts on the economy depending upon the number of workers involved, the sector affected, and the essentiality of the industry. A strike may affect an entire industry (e.g., coal miners), a giant company (e.g., General Motors), a local company or industry (e.g., Long Island Railroad), or only some workers in an industry (e.g., a construction industry local). Some strikes have been of extremely long duration (e.g., Kohler Co.); others have lasted for only a few hours (e.g., drawbridge operators in New York City). Some strikes may be terminated because of government action (e.g., national railroads), while others end because the parties decide on a new contract or on procedures to reach such an agreement.

Regardless of its size, each strike involves a cost: lost wages for workers,[11] lost profits to an industry, inconvenience to customers, and lost taxes to various governmental units.

Alternatives To Strikes

To avoid these costs several alternatives are available: mediation, fact-finding, and arbitration.[12] The Federal Mediation and Conciliation Service reports it helped settle 17,000 disputes a year in the 1968-71 period.[13]

Fact-finding involves the appointment of several impartial experts or observers to evaluate the arguments advanced by labor and management and to make recommendations which are advisory in nature. They help to mobilize public opinion behind the proposed settlement.

Table 1-5

MAN-DAYS IDLE THROUGH STRIKES,
BY YEARS, 1946–1972

	Number (Thousands)	Percent of Estimated Working Time
1946	116,000	1.04
1947	34,600	.30
1948	34,100	.28
1949	50,500	.44
1950	38,800	.33
1951	22,900	.18
1952	59,100	.48
1953	28,300	.22
1954	22,600	.18
1955	28,200	.22
1956	33,100	.24
1957	16,500	.12
1958	23,900	.18
1959	69,000	.50
1960	19,100	.14
1961	16,300	.11
1962	18,600	.13
1963	16,100	.11
1964	22,900	.15
1965	23,300	.15
1966	25,400	.15
1967	42,100	.25
1968	49,018	.28
1969	42,869	.24
1970	66,414	.37
1971	47,589	.26
1972	26,000	.14

Source: *Monthly Labor Review* (September 1973), p. 133.

Arbitration is widely used to resolve grievances and to a much lesser extent to determine the terms of a new contract. Arbitration takes a substantial amount of time, normally several months, takes on many aspects of a judicial procedure and involves substantial costs.

The growing concern over industrial peace has brought to the fore-front several proposals: greater activism by mediators to spur the parties to agreement, longer-term contracts,[14] "instant arbitration"—for grievances and mediation—and arbitration for the basic contract.

Instant Arbitration. One technique for resolving labor-management disputes peacefully and quickly is "instant arbitration," which is designed to resolve grievances without the necessity of a formal hearing. This approach is used at General Electric and in the steel industry. As Lawrence Stessin has pointed out, "this innovation in industrial peace is designed to keep employee grievances from exploding into wildcat strikes or degenerating into poor worker morale."[15] Thus, the testimony may take one day or less and the decision be rendered in a week.

Mediation-Arbitration. Another development is mediation-arbitration. In traditional mediation the function of the mediator is limited. He merely acts as a go-between in the hope of resolving the dispute. He does not have the power to prevent a strike. In conventional arbitration, the arbitrator holds formal hearings which involve the maintenance of a record, the testimony by witnesses, and the introduction of numerous exhibits. The decision rendered is binding on the parties, and it may have a significant impact on future bargaining.

On the other hand, in mediation-arbitration some sessions involve negotiations between the parties. The parties must be careful in the presentation of their positions because they are aware that the mediator-arbitrator ultimately may have to make the decision. As a result, the range of differences may be narrowed substantially and a settlement become possible. It is also likely that the process of mediation-arbitration may encourage more creative thinking by the participants and lead to greater flexibility in position than is found in traditional bargaining or mediation. This approach has been used in recent years in disputes involving the nurses in San Francisco, teachers in Hawaii, and the Pacific Coast longshore industry.

Which is the best procedure to follow? Out of his very rich experience, Mr. Cole concludes:

My strong preference is for an indefinite combination of all these procedures, together with any others the parties may devise that they believe may be helpful. I would favor a course in which they move by agreement step by step from one procedure to another until their differences are ended. I would exclude nothing, nor would I compel them to go forward if either chooses not to do so. This is because of the very nature of agreement making.

Although the approaches discussed above can contribute to a reduction in the number of strikes, as Mr. Cole concludes, "there is no formula or miracle device for avoiding labor strife." Mr. Cole has described the role of the National Commission for Industrial Peace as follows:

We should seek to induce in labor and management representatives an understanding of the purpose of collective bargaining and the disposition or desire to have it function effectively on a constructive, rather than destructive, basis. Without such a disposition, there is little that government can do to improve collective bargaining or to safeguard the public interest.[16]

The key to industrial peace is found in the attitudes of the parties rather than in mandates legislated by government.

BUSINESS AND TECHNOLOGY

New technology has made vital contributions to progress throughout our history. The major breakthroughs in such areas as communications, energy, and transportation provided important stimuli to our growth as a nation and to its great economic strength. The great growth industries of the post-World War II period—computers, ethical drugs, chemicals, electronics, synthetic fibers—have been frontiers of technological advances.

The energy crisis which had been casting its shadow for several years was thrust before a startled public by the Arab oil boycott late in 1973.

Here the role and the problems of technology quickly became evident in many of the proposed solutions: gasification of coal, nuclear energy, recovery of oil from shale, solar energy. The development of these alternative sources of energy require an intensification of research efforts and an enormous expenditure of funds, public and private.[17]

Earlier we had experienced other massive concentrations of research efforts to develop the technology required to produce the atomic bomb in World War II and for space exploration in the 1960s and early 1970s.

As the contribution of technology has been increasingly understood, American companies have devoted significant resources to planned research and development (R & D). Breakthroughs in technology often have been achieved by smaller firms (e.g., the oxygenation process in steel, and Xerography). But it is the larger companies that have the resources to mount ongoing larger scale expenditures for R & D to seek out new technology systematically.[18]

According to Dr. Drucker, innovation and technological change in the next quarter of a century "will have to be done in and by established organizations, and especially in and by established businesses." However, within such companies he notes that "Organizationally and managerially, technological activity still tends to be separated from the main work of the business and organized as a discrete and quite different R & D activity. . . ." He emphasizes that this relationship is unsatisfactory and that technology must be treated as a "central management task."

As background for these proposals, it is helpful to review the growth of R & D and its role in our economy.

Growth of R & D

Expenditures for research and development R & D have increased substantially from $6.3 billion in 1955 to $30.1 billion in 1973. Although part of this increase reflected general price inflation, there has been an enormous increase in real terms. Table 1-6 shows that total R & D increased much more rapidly than total GNP until the mid-1960s but is now rising at a slower rate.

Table 1-6

EXPENDITURES FOR RESEARCH AND DEVELOPMENT,
SELECTED YEARS, 1955–1973

Year	Gross National Product (billions)	Total R & D (millions)	Percent of GNP
1955	$ 398.0	$ 6,279	1.58
1960	503.7	13,730	2.73
1965	684.9	20,439	2.98
1970	977.1	26,566	2.72
1973	1,288.2	30,100	2.34

Sources: U.S. National Science Foundation and U.S. Department of Commerce.

More than half of total R &D has been financed by the federal government, particularly for armaments and space exploration. Increasingly, however, R & D is being financed by company funds:

PERCENT OF R & D FINANCED BY COMPANY FUNDS

1955	44.1
1960	36.3
1965	36.2
1970	44.4
1973	46.9

The number of scientists, engineers, and technicians in private industry and in government in 1970 was almost 1.9 million as compared with total nonagricultural employment of 70.6 million. According to the National Science Foundation, the full-time equivalent number of research and development scientists and engineers increased from 229,400 in January 1957 to 387,100 in January 1969 and then declined to 359,300 in January 1971.[19]

R & D and Economic Growth

Since World War II, economic growth has been a major objective of economic policy here and abroad. Growth has been necessary to provide jobs for our expanding labor force, to make possible higher levels of living, and to meet the expectations of minority groups. Moreover, the expanding role of government could be supported only by a rising national output if the average level of living for the population was not to be reduced.

The ultimate source of economic growth is found in our willingness to forgo consumption today so that the capital investment can be created to expand future production. This investment is made in plant and equipment, and in intangible capital (health, education and training, research and development).[20] The development of new technology is particularly important because it expands the quantity of usable resources, results in production economies, and adds to diversity of goods and services available.

Product Differentiation and R & D

With some exceptions, producers seek to build unique qualities and services into their products in order to induce consumer purchases. The resulting product differentiation contributes to effective competition. The wide diversity of models for automobiles and appliances, and the development of different drug formulations to meet the same illness provide familiar illustrations. These differences are stressed in advertising and sales campaigns by producers who seek to provide that "little something extra" which will appeal to the consumer. The result is competitive pressure for R & D to develop product improvements or modifications which will result in product differentiation and add to sales appeal. It must be recognized, of course, that some product modifications are only frills which may have questionable value.

Competition and R & D

For many products competition in terms of price is subordinated to nonprice factors including service, delivery, credit terms, quality, novelty, and others. The most important form of competition, as Professor Joseph Schumpeter noted, is, "the competition from the new commodity, the new technology, the new source of supply, the new type of organization . . . competition which commands a decisive cost or quality advantage and which strikes not at the margins of the profits and the outputs of existing firms but at their foundations and at their very lives."[21]

This is exactly the area in which R & D plays its most important role. An improvement in competitive position provides the strongest incentive for firms to make R & D expenditures.

Role of Business

It is Dr. Drucker's thesis that "technology can be anticipated" and that "business managers have to learn that technology is managerial opportunity and managerial responsibility." The businessman's task is "to understand what new knowledge is becoming acceptable and available, to assess its possible technological impact, and to go to work on converting it into technology—that is into processes and products." Thus, the major role of business is to convert, that is to develop, promising ideas into useful and usable products and processes.[22]

The frontiers of technology are still wide open. Those who refuse to be shackled by the past will find ample opportunities to develop these new technologies as new problems arise. The organizational approach advocated by Dr. Drucker will facilitate achieving this objective.

PRODUCTIVITY AND BUSINESS

Increases in productivity play a key role in the economic well-being of all groups in our economy:

- For business, it is a means of cutting costs and hence may contribute to higher profits.
- For labor, it is the source for higher wages and fringe benefits, including the shorter work week.
- For government, it helps to increase the tax base and thus to finance expanding programs of activity.
- For investors, it provides opportunities for profitable use of savings.
- For consumers, it may make possible relatively lower prices for some goods and services.
- For the population generally, it is the source of higher levels of living which are made possible when total output grows more rapidly than population.

Productivity measures the relationship between all of the inputs of resources—labor, materials, capital, management—and the resulting output. New investment in tangible capital, that is, plant and equipment, is one of the most important factors contributing to higher productivity. As the Council of Economic Advisers has noted: "To keep output per worker rising rapidly, when the labor force is also rising rapidly, requires a high rate of investment in productive facilities."[23] However, contributions also are made by significant " 'intangible investments' in research and development, education, training, health, and mobility, all of which raise the quality of the factors in which the resulting intangible capital is embodied."[24] Another factor which may add to productivity is the scale of production. As Dr. Kendrick notes, high volume provides a favorable background for productivity gains and vice versa—particularly in short periods.

The role of management also must be emphasized. Production planning contributes to total productivity by the manner in which plants are laid out, adequate inventories are available, sources of raw materials are assured, training and educational programs are established, the speed with which new equipment and new processes are introduced, and so on.

Although productivity reflects the contributions of all of these factors of production, data usually are available on a continuous basis only in terms of output per man-hour (OPM) (total production divided by some measure of man-hours) or output per employee (total production divided by the number of employees).[25] Dr. Kendrick describes these measures as "partial productivity ratios."

OPM shows the total physical volume of output (arising from the combination of manpower, tool power, materials, etc.) as compared with one form of input, namely, the hours worked or paid for or the number of employees. Man-hours or employees are used because the data are more readily available, are more easily measured, and are common to all industry. Thus, an increase in OPM will overstate the rise in efficiency if the savings in man-hours result in part from the substitution of capital for labor. However, because of the difficulty in obtaining satisfactory output data, particularly for industries with widely diversified products whose nature and quality change markedly over time, meaningful OPM data cannot be derived for all industries.

Productivity and Labor Costs

In recent years, the relationship between changes in wages and in output per man-hour (OPM) has been given increasing attention. Impetus was given to the use of this criterion by the inclusion of an annual improvement factor in the General Motors Corporation contract in 1948.[26] Some years later the President's Council of Economic Advisers set forth guideposts for wages and prices based upon changes in OPM. Thus, in the 1962 Economic Report, the Council noted that in any industry "the annual rate of increase of total employee compensation (wages and fringe benefits) per manhour worked should equal the national trend rate of increase in output per manhour."

The importance of the productivity criterion as a test of the appropriateness of annual wage adjustments should not be overemphasized.

There has been no period in which wages and productivity were in perfect balance. Productivity is only one of many factors that enter into wage determination.

Comparisons often are made between changes in average hourly earnings and output per man-hour. Because of the widespread introduction of nonwage benefits, average hourly earnings do not provide a basis for satisfactory comparison. OPM gains may be used to pay higher wages and/or improved nonwage benefits. The comparison between OPM and total labor costs is useful as a measure of unit labor costs (ULC).

Other Claimants

In addition to labor, there are other claimants to productivity gains. Those providing the savings must be rewarded, government expects its share in taxes, and consumers can share by obtaining relatively lower prices. When a product's price falls in relation to other prices, effective demand tends to increase. Distribution of the gains in productivity through lower prices is particularly important because it creates additional employment and investment opportunities in the industry affected and acts to prevent unemployment.

Considerable emphasis has been given in recent years to the desirability of increasing productivity as a means of containing inflation. For example, in June 1970 the President established a National Commission on Productivity, which had as one objective stimulating OPM as an offset to the increases in hourly labor costs.[27] However, the experience to date suggests that the Commission has not been very successful in this area.

The Record of Output per Man-hour

Historically, output per man-hour for the private economy has increased at an average annual rate of about 3 percent. However, the annual rates of changes have varied widely from this average. (See Table 1-1.) In some years the increase has been less than 1 percent (1956,

1969), while in others it has exceeded 4 percent (1948, 1950, 1955, 1962, 1971).[28]

The record for major industries has been very diverse. Between 1960 and 1972, for example, the annual rate of increase in OPM for all employees was 6 percent or higher for malt liquors, man-made fibers, radio and television receiving sets, petroleum pipelines, and air transportation and less than 2 percent for footwear, primary copper, lead, and zinc, and cigarettes.[29] (See Table 1-7.)

There has been a slowdown in the rate of increase in productivity since 1966. Dr. Kendrick attributes this development to several factors: "negative social tendencies" including the "critical attitude toward economic growth and materialism" and "some weakening of the work ethic," actions by workers to "resist technological innovations," price inflation, "inadequate profit margins" which have led to "materials shortages," "significant increases in antipollution outlays," change in labor-force mix with a larger percentage of women and youngsters, "direct governmental intervention in markets," the "total tax burden," and the "cut-back in government financed R & D."

Future Productivity Gains

Future national gains in productivity, however measured, will be determined in part by the policies we adopt and in part by the changing structure of the economy.[30] Dr. Kendrick notes there is "a whole menu of policy measures" which could "bolster productivity advance." He underlines the importance of providing inducements for greater private expenditures for R & D because of "Its key role in productivity advance." The inducement would take the form of an investment tax credit with the greatest incentives to manufacturers of producers goods. He also proposes "a policy of steady growth" in federal funding of R & D to avoid such sharp declines as that between 1969 and 1972 "with the attendant sharp increases in unemployment of scientists and engineers."

Table 1-7

SELECTED INDUSTRIES: AVERAGE ANNUAL RATES OF CHANGE IN
OUTPUT PER MAN-HOUR, ALL EMPLOYEES, 1947–1972 AND 1960–1972

| | Output per Man-Hour (Average Annual Percent Change)[1] | |
	1947–72	1960–72
Petroleum pipelines	NA	9.6
Air transportation	7.7	7.8
Radio and television receiving sets	NA	6.5
Hosiery	5.3	6.4
Malt liquors	5.1	6.3
Man-made fibers	NA	6.0
Gas and electric utilities	6.8	5.8
Bottled and canned soft drinks	NA	5.5
Class I Railroads, revenue traffic	5.2	5.5
Petroleum refining	5.7	5.3
Aluminum rolling and drawing	NA	5.3
Major household appliances	NA	5.0
Concrete products	3.5[2]	4.7[3]
Flour and other grain mill products	4.0	4.3
Paper, paperboard, and pulp mills	4.0	4.3
Hydraulic cement	4.5	4.2
Cigars	5.8	3.8
Sugar	4.3	3.7
Corrugated and solid fiber boxes	NA	3.5
Tires and inner tubes	4.0	3.4
Motor vehicles and equipment	NA	3.3
Class I Railroads, car-miles	4.2	3.3
Bakery products	2.4	3.2
Primary aluminum	4.7	3.1
Candy and other confectionery products	3.2	2.8

| | Output per Man-Hour (Average Annual Percent Change) | |
	1947–72	1960–72
Glass containers	1.7	2.5
Gray iron foundries	2.2[4]	2.4
Canning and preserving	3.3[2]	2.3[3]
Ready-mixed concrete	NA	2.2[3]
Steel	1.7	2.2
Tobacco products—total	3.2	2.1
Metal cans	2.4	2.0
Primary copper, lead, and zinc	2.2	1.7
Cigarettes, chewing and smoking tobacco	1.4	1.2
Steel foundries	1.3[4]	1.2
Footwear	1.4	0.5

NA = Not available.
1. Based on the linear least squares trend of the logarithms of the index numbers.
2. Average annual rate of change is for 1947–71.
3. Average annual rate of change is for 1960–71.
4. Average annual rate of change is for 1954–72.

Source: U. S. Department of Labor, Bureau of Labor Statistics, "Indexes of Output Per Man-Hour: Selected Industries, 1973," *Bulletin 1780* (Washington, D.C., 1973), p. 3.

The adverse impact of price inflation on productivity could be minimized by the adoption of "some form of inflation accounting." According to Dr. Kendrick, absent such a change, the funds available for new investments are eroded because part of retained earnings must be used to meet higher replacement costs for inventories and plant and equipment. Realistic accounting for inflation would result in lower taxes on profits, thus "increasing internal sources of funds."

While the above proposals could have a favorable impact on productivity, structural changes in the economy will have a negative impact on national productivity. It is probable that future gains in OPM will be somewhat lower for the private economy than it was between 1948 and 1969. Two factors point in this direction: (1) increasing relative importance of services, and (2) the probable lower rate of future growth in real gross national product.

Satisfactory data are not available for OPM for all services. But the available evidence indicates that the average rates of increase in OPM in the service industries have been significantly lower than for agriculture and industry.[31] Dr. Kendrick's data show that real productivity per man-hour increased only 1.1 percent per year from 1948 to 1969 as compared with 3.1 percent for the private domestic economy. (See Table 5-1.) With the service industries accounting for an increasing proportion of our national economic activity, their lower rates of increase in OPM will act to depress the average rate of gain in OPM for the economy.

Since OPM is an important component of economic growth, smaller increases in total OPM will have an adverse effect upon long-term growth rates. Increases in total output provide a favorable background for greater OPM because of the opportunities created to utilize more tangible and intangible investment. Our economic growth has been facilitated by ample supplies of low-cost energy. It has become increasingly clear that energy costs will rise substantially and that, absent a major technological breakthrough, supplies of energy will increase at lower rates than in the past. In addition, the worldwide shortages of basic raw materials in 1973-74 have raised serious questions concerning the extent to which supplies available to this country will increase in the future. These developments are bound to be accompanied by lower rates of growth in real GNP and thus to provide less stimulus to greater productivity than in the past.

NOTES

1. John W. Kendrick, *Productivity Trends in the United States* (Princeton: Princeton University Press, 1961), and *Postwar Productivity Trends in the United States, 1948-1969* (New York: National Bureau of Economic Research, 1973).

2. Jules Backman, *Wage Determination* (Princeton: D. Van Nostrand, 1959), Ch. 10.

3. These include: Norris-LaGuardia Act, 1932; The National Industrial Recovery Act, Section 7a, 1933; and The National Labor Relations Act (Wagner Act), 1935.

4. Jack Stieber, *Public Employee Unionism: Structure, Growth, Policy* (Washington: The Brookings Institution, 1973).

5. Wage supplements which cover primarily pension, welfare funds, unemployment compensation and social security have increased from 4.6 percent of total labor compensation in 1947 to 11.3 percent in 1972.

6. As part of this agreement, national strikes and lockouts are prohibited, workers were guaranteed wage increases of at least 3 percent a year in August 1974, 1975, and 1976, the escalation clauses were continued, and a bonus of $150 was to be paid to each employee. *Monthly Labor Review* (May 1973), p. 62.

7. A. H. Raskin has reported that "many unions have concluded that there is little percentage in calling long strikes for more money when the money loses value so fast." *The New York Times,* February 17, 1974, Sec. 4, p. 2.

8. David Larson and Lena W. Bolton, "Calendar of Wage Increases and Negotiations For 1973," *Monthly Labor Review* (January 1973), p. 8.

9. *Social Security Bulletin* (March 1973), p. 4.

10. In the public sector, there was an average of 350 strikes a year between 1968 and 1972 with an average of about 1,500,000 man-days lost. *The Morgan Guaranty Survey* (February 1974), p. 10.

11. In a seasonal industry such as coal, a short strike in off-season periods may have no adverse effect upon the total annual wages of coal miners in that year.

12. For a brief discussion, see A. L. Gitlow, *Labor and Industrial Society*, rev. ed. (Homewood, Ill.: Richard D. Irwin, Inc., 1963), pp. 454-57.

13. Federal Mediation and Conciliation Service, *Twenty-Fourth Annual Report Fiscal Year 1971* (Washington, 1972), p. 28.

14. This creates a problem during periods of price inflation unless provision is made for automatic cost of living adjustments or for periodic reopening of the contract to consider increases in living costs.

15. Lawrence Stessin, "Speed is the Essence," *The New York Times,* February 17, 1974, Sec. 2, p. 2.

16. David L. Cole, "The Search For Industrial Peace," *Monthly Labor Review* (September 1973), p. 39.

17. Henry Kissinger told the United Nations that the United States was

allocating $12 billion for energy research and development over a five-year period. *The New York Times,* April 16, 1974, p. 12.

18. See testimony of Jesse Markham in *Hearings on Economic Concentration* before the Subcommittee on Antitrust and Monopoly, Committee on the Judiciary, United States Senate (Washington, 1965), Part 3, p. 1276; A.D.H. Kaplan, *Big Enterprise in a Competitive System* (Washington: The Brookings Institution, 1964), Ch. 10; *Economic Report of the President,* January 1972 (Washington, 1972), pp. 125-30.

19. *Research and Development in Industry, 1970,* NSF 72-309 (Washington: National Science Foundation, 1972), p. 58.

20. R. Nelson, Merton J. Peck, and Edward D. Kalachek, *Technology, Finance, Growth, and Public Policy* (Washington: The Brookings Institution, 1967), and Jacob Schmookler, *Invention and Economic Growth* (Cambridge, Mass.: Harvard University Press, 1966).

21. Joseph A. Schumpeter, *Capitalism, Socialism and Democracy,* 3rd ed. (New York: Harper, 1942), p. 84.

22. For a brief analysis of the problems of forecasting new technology see Oskar Morgenstern, Klaus Knorr, and Klaus P. Heiss, *Long Term Projections of Power* (Cambridge, Mass.: Ballinger Publishing Co., 1973).

23. *Economic Report of the President,* February 1974 (Washington: Council of Economic Advisers, 1974), p. 37.

24. Kendrick, *"Postwar Productivity Trends in the United States," op. cit.,* p. 3.

25. OPM may be measured in terms of production workers, nonproduction workers, or total employees. See "Indexes of Output Per Manhour, Selected Industries 1973 Edition," *Bulletin 1780* (Washington: U.S. Bureau of Labor Statistics, 1973).

26. For a detailed analysis of the General Motors approach, see Backman, *op. cit.,* Ch. 9.

27. *Economic Report of the President,* February 1971 (Washington: Council of Economic Advisers, February 1971), p. 91.

28. *Economic Report of the President,* February 1974 (Washington: Council of Economic Advisers, 1974), p. 287.

29. "Indexes of Output Per Manhour, Selected Industries," *op. cit.,* p. 3.

30. *Conference on an Agenda for Economic Research on Productivity* (Washington: National Commission on Productivity, April 1973).

31. Victor R. Fuchs, *The Service Economy* (New York: National Bureau of Economic Research, 1968), Ch. 3, and *Production and Productivity in the Service Industries,* Victor R. Fuchs, ed. (New York: National Bureau of Economic Research, 1969).

TWO

Labor in Transition

Emanuel Stein

*Professor of Economics, Humanities
and Social Sciences
New York University*

In an assessment of the labor movement—where it is and where it seems to be going—it is essential to keep in mind some fundamentals bearing on the nature and composition of unions, their goals and tactics and interunion relations. *First*, the predominant characteristic, as R. F. Hoxie pointed out over half a century ago,[1] is *business unionism*, with its focus on economic gains for the constituency chiefly through collective bargaining. Nearly all American unions—certainly those that have lasted—are of this type. Moreover, the attractiveness of a union to actual and prospective members depends largely upon the ability of the union to "deliver the goods" in wages and job security. Success in collective bargaining depends, in turn, upon the ability of the union to achieve sufficient control of the labor supply or the job market or both to exert effective pressure upon the employer. The structure of the union and the pressure for organization are, thus, mainly determined by the requisites for effective bargaining.

Second, and as a direct result of the primacy of business unionism, the labor movement is a very decentralized affair. The AFL-CIO is, in fact as well as in name, a loose federation of national unions over whom the central body has little control. In a public relations sense, the AFL-CIO speaks for the labor movement as a whole and, depending upon the nature of the issue and the vigor of its officers, can exert some moral suasion upon its affiliates, especially the smaller ones. But its ultimate sanction of expulsion rarely is effective against the large affiliates which, time and again, have demonstrated their ability to secede from the Federation without any impairment of strength.

In this sense, the Federation resembles a congress of sovereign states—the national unions—more or less warmly disposed toward each other, having some ill-defined and vague interests in common, and capable of cooperating effectively from time to time, but jealous of their sovereign status and quick to react against threats to that sovereignty. The AFL was founded upon this idea of trade autonomy and its twin idea of exclusive jurisdiction. Notwithstanding the difference in the rhetoric, when the CIO was a separate federation it, too, operated upon the basis of substantially autonomous affiliates influenced, but not controlled, by the federation. In recent years, there has been some small accretion of power to the AFL-CIO and a somewhat greater effective influence over the affiliates than in the days of Samuel Gompers or William Green. But basically the national unions go their own way, determining their own goals and policies free of centralized direction and control.

Third, the labor movement has rather prided itself on a refusal to define ultimate objectives, on a rejection of theoretical formulations, and on an exclusive concern with the "immediate." Samuel Gompers put it thus:

> In improving conditions from day to day the organized labor movement has no "fixed program" for human progress. If you start out with a program everything must conform to it. With theorists, if facts do not conform to their theories, then so much the worse for the facts. Their declarations of theories and actions refuse to be

hampered by facts. We do not set any particular standard, but work for the best possible conditions immediately obtainable for the workers. When they are obtained then we strive for better.[2]

The constant objective or goal has been "more, now"—as continuous and substantial betterment of the economic conditions of the members as times and circumstances would permit. The approach has been pragmatic; it seeks to operate within the system of private profit-seeking enterprise, though on occasion, under special conditions, it has advocated public ownership. Radical changes in the economic and political spheres have been strenuously opposed;[3] it is to be noted that, in the early 1970s, left-wing unionism had virtually ceased to exist. The labor movement may thus be regarded as essentially conservative in its outlook, confining its activities mainly to collective bargaining. Here and there, a union may sponsor cooperative housing, cooperative buying, or some other welfare activity for its membership. But the main thrust has been, and continues to be, on the members' interests as employees, on their wages and conditions of employment.

Fourth. Notwithstanding the emphasis upon collective bargaining, but partly as a result of it, the labor movement's relations to, and dependence upon, government have become very close and continuous. For many years, unions demonstrated antipathy to labor legislation, desiring basically to be left free to work out their destiny on the economic front. Thus, they sought relief from restrictions on strikes, pickets, and boycotts, and fought for the elimination of the labor injunction and for exemption from the antitrust laws.[4] On the side of protective legislation, the Federation favored laws against women's and child labor, prison labor, and immigration, but opposed wage-and-hour laws for men and unemployment insurance. The advent of the post-1929 depression and the inauguration of President Roosevelt produced a profound alteration in these historic attitudes. Labor began to look to government—especially the federal government—for the direct establishment and improvement of labor standards and for *indirect* help by strengthening unions on the collective-bargaining front. Among the immediate results, in the prevailingly sympathetic climate of public and official opinion, were such laws as the National Industrial Recovery Act, the National Labor Relations (Wagner) Act (NLRA), the Social

Security Act, the Public Contracts (Walsh-Healy) Act, and the Fair Labor Standards (Wage and Hour) Act. Decisions by the Supreme Court, overruling old cases and revolutionizing traditional interpretations of the commerce and police powers, affirmed the constitutionality of the new legislation.[5] The way was then cleared for unprecedented extension of state and federal action on labor matters.[6]

Involvement with government was necessarily vastly increased during World War II, as the labor movement participated in a broad range of wartime activities, including the War Manpower Commission and the National War Labor Board. Following World War II, there was no abatement of governmental activity vis-à-vis labor, and, hence, no diminution of labor's concern with what was going on at the governmental level: e.g., the Labor Management Relations (Taft-Hartley) Act, the Labor Management Reporting and Disclosure (Landrum-Griffin) Act, and the Civil Rights Act of 1964. Evidently, government could be a powerful ally; it could also be a formidable antagonist. Understandably, there was great concern within the labor movement that government should be the one, and not the other.

UNION MEMBERSHIP

There has always been a good deal of uncertainty about the total of union membership in the United States, attributable in part to the reluctance of some unions to disclose their membership, the difficulty of locating small unaffiliated unions, doubts as to whether particular organizations (especially of professional persons) should be classed as unions, and diverse standards for classifying individuals as members. Textbooks on labor typically furnish tables on AFL-CIO membership from the establishment of the AFL in the 1880s, leaving at least the implication that the strength and well-being of American unions are to be measured in this way. Expressed as a fraction of the total labor supply, it appears that union membership in the private sector of the economy has suffered a substantial decline since the end of World War II.

In recent years, the labor movement has come under sharp attack from some liberals (including prominent union officials) who, pointing to the fact that only a little more than one quarter of the nonagricultural working force belongs to unions, charge the unions with failure to meet their responsibilities and opportunities. Criticism from this source has usually focused on the absence of significant efforts to organize agricultural, white-collar, and service workers, as well as other groups theoretically organizable. Unions are accused of having become smug and contented with their successes at "bread-and-butter" or business unionism, instead of setting broad national goals of social betterment. It is also asserted that the idealism which characterized the great organizing drives of the 1930s has been replaced by a bland self-centered materialism devoted solely to bargaining.

Obviously, questions as to the social responsibilities of unions and of particular philosophies to be pursued are matters on which reasonable men may disagree, and those who think that contemporary unions should play down collective bargaining and address themselves to more comprehensive social programs are entitled to their views. But it is a plain misreading of American labor history to suggest either that the organizing activities have been diverted from their traditional channels or that all that stands in the way of enormously expanded union membership is a willingness of unions to organize the unorganized. Further, it is a mistake to equate overall union membership with union power as expressed in collective bargaining. And, in any event, there are sound reasons for doubting that a great organizing effort will be made, or, if made, will meet with any great success.

First, the structure of American unionism is not conducive to the comprehensive organizing programs envisioned by labor's critics. Each of the national unions has its own sphere or spheres of interest—the industries or the jobs in which its members are engaged. To the extent that the economic interests of its members would be advanced by organizing some particular groups of workers in such industry or in such skills, it is to be expected that the requisite organizing effort will be forthcoming. On the other hand, there is no reason to suppose that unions would be willing to expend the large sums required to organize workers who are in an unrelated industry or occupation, where the "benefits" from enlarged membership would accrue to some other union. There are many uses to which funds can be put which would provide more immediate benefits for the membership, and unions are not indifferent to such considerations. It would obviously make more sense for a large organizing campaign to be made by the AFL-CIO, but the Federation itself is dependent upon the payments made to it by its affiliates, so that the burden of cost would still fall upon the latter.

Second. The experience of the 1930s was perhaps the most unique in American labor history. The newly organized CIO did indeed mount large, expensive, and successful campaigns. On the other hand, the country was, so to speak, ripe for organization, with hundreds of thousands of workers rushing to join unions. There is nothing whatever to suggest any such climate of opinion at the present time. Experience in NLRB certification elections over the past decade does not indicate any overpowering demand by unorganized workers for organization. For better or worse, persons who join unions do so because of belief or hope that their circumstances will be improved thereby, and it is clearly true that the only really successful appeal unions can make in recruiting campaigns is to the self-interest of the prospective members. This, it may be observed, is no less the case with public employees or college professors than with manual workers. In a period when sophisticated employers are aware of the temptation of their unorganized employees to seek the benefits of unionization, there is less room for union organizers to appeal to the employee's self-interest.

Third. Where employees are substantially ripe for union organization, the provisions of the Labor Act and the facilities of the NLRB (the prohibitions of employer unfair labor practices specified in Section 8 [a], and provisions for elections and certification) facilitate the task of the union. On the other hand, it is no longer possible to organize "from the top," organizational picketing is subject to limitation, and the union must be able to demonstrate a substantial following among the employees in order to be entitled to a representation election. In such a situation, organizing efforts are apt to become more time-consuming and expensive.

Fourth. Here and there, one finds special cases of organizing effort in which a union seeks to extend its power and prestige by entering a field outside its customary sphere. Prior to the merger of the AFL and the CIO and prior to the negotiation of the no-raiding pacts between them, there was a good bit of reciprocal raiding. For the most part, such activities had little effect upon overall union membership, since a successful raid meant merely the transfer of an existing bargaining unit from one union to another. In the process, it sometimes happened that hitherto-unorganized establishments were brought within the unionized sector, but it is to be doubted that the motivation here was the furtherance of the kind of social program referred to above.

Fifth. Over the past several years, there has been a large gain in union membership in the public sector, both federal and local. Organization here has come almost entirely from within, as the public employees across the board rushed to join existing public service unions or to form new ones. The pressure for economic improvement, such as had been achieved in the private sector, was strong. Further, the process was helped by the adoption of legal and administrative measures [7] which removed many of the former barriers to organization, expanded the scope of bargaining,[8] and provided machinery for the adjustment of disputes.[9] The legal obstacles to organization of public service employees had in many instances been formidable, notwithstanding that many private-sector unions counted government employees among their members; there were also the problems presented by the circumscribing effects of civil service laws on the area of permissible union activity. These proved insufficient to overcome the pressure exerted by employees insistent upon self-organization and collective bargaining.

Sixth. The experience in the public sector—that traditionally nonunion employees are susceptible to effective organization if the will is there—is underscored by developments in other areas. Thus, resident physicians in New York City hospitals are organized into a Council of Interns and Residents and are engaged in active collective bargaining. Professional athletes in a variety of sports are similarly engaged, and progress has been made in recent years in the organization for bargaining purposes of college and university faculties. There is no reason to suppose that traditional views of the nonorganizability of such groups as white-collar and service workers are really valid any longer. Accordingly, one must conclude that, if the times and circumstances are sufficiently favorable to organization, unionization will occur here, as it has elsewhere, with or without mass organizing campaigns.

On balance, it does not now seem likely that there will be any striking changes in overall union membership in the United States. There is nothing to indicate that the labor movement is prepared to launch the major drive which would be required to organize those industries now largely unorganized, or that, if launched, it would be successful. Nor do the individual unions, which traditionally have the main responsibility for organization, manifest a purpose to organize the hitherto-unorganized segments of their industries on a broad basis. It may be worth noting that about 9000 certification elections were conducted by NLRB in fiscal 1972 affecting about 570,000 employees, that the average number of employees voting per establishment was 58 (as compared to 62 in fiscal 1971), and that unions won 55 percent of the elections, somewhat more than in the preceding year; 86.2 percent of the elections involved fewer than 100 employees each.[10] It is also interesting to note the narrow range of the election results over recent years. In the elections conducted by the Board in the period 1962-72, unions have been successful in about 57 percent of the cases with a range of from 54 to 62 percent.[11] These figures suggest no great organizing activity. Moreover, they represent a continuation of the fairly persistent, though moderate, decline of representation elections as a part of the work of the NLRB. One may hazard the guess that it would take a major change in the economic climate and in public opinion to produce a marked increase either in organizing activity by unions or in the desire of the unorganized to join.

Several other points must be made in connection with the matter of membership. First, the ability of a union to bargain successfully for its members depends upon *its* ability to organize *its* markets, not upon the overall extent of union organization. While a growth in total AFL-CIO membership may serve a variety of purposes—especially, the enhancement of political influence—it is of relatively small importance in the vital area of collective bargaining. From this standpoint, it is rather pointless to emphasize the AFL-CIO membership totals, for the AFL-CIO does not engage in collective bargaining. When the chips are really down, the crucial question is the strength of the individual union, not the size of the labor movement. Thirty million members scattered over the United States would be far less significant than half that number concentrated in key industries or areas. In this respect, there is a startling contrast between the present situation and that of the early 1930s. At that time, union organization was concentrated in but a few industries: coal mining, construction, printing, the needle trades, and railroad transportation; in the mass-production industries, unionism was conspicuously absent. At the present time, unions are firmly entrenched in the basic industries: steel, aluminum, automobiles, mining, electrical equipment, transportation, public utilities (including telephones), paper and paper products—to mention only a few. So far as impact on the total economy is concerned, it is obvious that the union strength in our most basic industries is of vastly more importance than would be its strength (as measured by an equal number of members) in less critical industries.

Moreover, a union's strength may be either greater or smaller than suggested by the size of its membership. Under the National Labor Relations Act, the selection of the bargaining representative is in the hands of a majority of the employees in the appropriate bargaining unit; unions may be chosen as bargaining representatives, and often are, though they do not have a majority of the employees in the unit as union members. The influence of the unions is in many instances, therefore, considerably wider than the size of the union membership would suggest, though there are many instances in which unions are not chosen as bargaining representatives despite a substantial membership among the employees. On balance, it is almost certainly true that union bargaining strength is significantly supplemented, rather than diminished, by nonunionists in the bargaining unit.

ROLE OF COLLECTIVE BARGAINING

Section 8(a)(5) and 8 (b)(3) of the National Labor Relations Act impose upon the employer and the bargaining representative of the employees respectively the duty to bargain collectively, as that term is defined in Section 8(d). Litigation over the meaning and application of the statutory provisions has resulted in an enormous case literature dealing with such matters as the determination of the bargaining representative, the procedural steps required to satisfy the duty to bargain, mandatory subjects of bargaining, optional subjects, subjects about which the parties are forbidden to bargain, and remedies which may be prescribed by the Board where it has found a statutory violation. For the most part, the scope of the duty to bargain has long been defined and delineated in Board and court decisions, so that the law on the subject may be regarded as fairly well settled. From time to time, however, new questions emerge. For example: Is an employer obliged to bargain over benefits to retired employees? Does a collective-bargaining agreement survive, in whole or part, the sale or merger of a business? What are the bargaining obligations of successors, and under what conditions may an employer withdraw from a multiemployer bargaining unit?

Thus, the process of "fleshing out" the law—filling in the gaps—continues, though at a moderate pace and primarily in respect to application of established principles of law to novel situations. Yet, a comparison of the law as it now stands with the law as it stood a quarter of a century or so ago reveals that there has been an extraordinary expansion of the mandatory subjects of bargaining and that today there are few matters, arguably within the scope of wages, hours, and terms and conditions of employment which are not required to be bargained about under the law.

As one proceeds from the legal aspects of collective bargaining to the everyday labor-management relations, one is struck by a marked and widespread rank-and-file restiveness whose influence upon unions and collective bargaining developments, though uncertain, may be of first-rate importance. This restiveness may be manifested in part by an increase in the number of unfair labor practice charges filed with the NLRB; in fiscal 1972, 26,852 such charges were filed—an all-time high—as compared with 23,770 in 1971 and only 13,479 in 1962.[12] In part, it may be manifested in legal actions, both private and other, under the several titles of the Labor-Management Reporting and Disclosure Act (Landrum-Griffin).[13] In many situations, it is expressed as an increase in the number of grievances processed. In still others, it finds expression in the refusal to ratify union-management agreements reached in collective bargaining. Moreover, it may take the form of contested union elections and the displacement of union officials.

It is not clear to what extent this is a purely temporary phenomenon, though assuredly contemporary events have provided much fuel for the flame. There can be little doubt, however, that there is here some expression of more durable discontent and of demand for change. It is important to observe that typically there is no antiunion animus here, for the characteristic attitude is one of loyalty to the institution. Rather, the hostility is directed against the leaders who are charged with dereliction of duty, if not worse. Members have always looked to the leaders to bring home "the bacon" from collective bargaining. Under the pressure of inflationary rises in the cost of living, the expectations and the felt needs have risen, and disappointments with the results of negotiations have become more frequent. Consequently, ratification votes often have become the occasions for bitter attacks upon the officers, charges of treachery and "selling the members down the river" are commonplace, and rejections of negotiated agreements tend to be a not-unusual reaction of the members. Often whole segments of a broad geographic bargaining unit resist a settlement even when it has been agreed to by the membership, as in recent strikes by the postal workers and the telephone workers in New York. In any event, the resentments are likely to persist.

It has been suggested that the restiveness is attributable in part to the attitudes of younger workers who are better educated than their seniors and more prone to challenge authority. Their outlook on life and about their jobs is said to be different from, and more militant than, that of older men. Obviously, this sort of thing does not lend itself to scientific verification, but one's personal experience in a very broad segment of American industry suggests that more recent arrivals in the labor force do react more vigorously and frequently to management and union conduct of which they disapprove and are quicker to resort to what they consider necessary remedial action.

Whatever the cause, there can be no doubt that many unions are now characterized by unusual intransigence on the part of the membership. There is a greater willingness to strike, notwithstanding that strikes are often averted by last-minute settlements, and a greater willingness to defy authority—not alone the authority of the employer or of the union, but the authority of the government (including the courts) as well. Disobedience of injunctions against strike action is no longer a rarity, nor the fining or imprisonment of union leaders for such disobedience.

Thus, teacher union leaders in New York, Newark, and Philadelphia have been sentenced to jail for violating injunctions, as has the leader of the New York sanitation workers. Nothwithstanding prohibitions by law of strikes by public employees, such strikes have occurred in growing numbers, so that the existence of the prohibitions appears to be more a minor inconvenience than an effective deterrent.

Militancy of the membership translates itself quickly into militancy of the leadership. One does not need to look far for examples of union officers who have been ousted from their jobs by a dissatisfied constituency. If the officers are to hold their jobs, they must be responsive to an articulate membership, and the most important test of their performance lies in the bacon they bring home from the bargaining. In the circumstances of these times, the officers' tenure is threatened far less by an intransigent bargaining posture than by a quick settlement of contract terms. It is to be expected, therefore, that the bargaining positions assumed by union negotiators are markedly stiffened and that they will insist on their demands even when it is highly probable, if not certain, that these demands, if agreed to, would not be approved by governmental control agencies when wage control is in effect. This was illustrated in a number of situations in 1973-74.

Union leaders are certainly aware of the interrelationships between labor costs and unemployment. There are occasional instances in which the threat of large-scale unemployment is so real and immediate as to persuade the union to make substantial concessions in order to save jobs.[14] To an overwhelming extent, however, the prospect of some unemployment, even if believed, is not permitted to detract from the pressure for wage increases; the trade-off is jobs for money. This is understandable, since those who have jobs—and will presumably continue to have them, so that they will benefit from wage increases—are in the ordinary case more numerous (and more influential in the union) than those who may lose their jobs as a result of increased labor costs. The felt need for more money is continuous and great, especially when living costs are rising sharply, and there is no prospect of abatement in the resulting pressure in the foreseeable future, though unemployment should continue to mount.

The drive for higher wages has been, and is likely to continue to be, accompanied—and at an accelerated pace—by the drive for expanded fringe benefits. From the quite modest beginnings during World War II, when the National War Labor Board approved three paid holidays a year and vacations of one week after one year's service and of two weeks after five years' service, there has been a truly amazing expansion in both the kinds and levels of fringe benefits: holidays, vacations, pensions, paid sick leave, leave for death in the family, pay while on jury duty or for military encampment, etc. The varieties of fringe benefits seem to be coextensive with human ingenuity. In the field of health care, for example, the original rather simple basic medical and surgical plans have in recent years been supplemented by dental plans, drug plans, visual care benefits, benefits for psychiatric treatment, health checks and screening benefits, to say nothing of major-medical plans.[15] The Social Security Administration estimates that contributions for employee health and insurance benefits more than doubled between 1965 and 1971, amounting in the latter year to more than $20 billion.[16] An interesting, and potentially very significant, development in fringe plans is to be seen in the 1974 agreement in the aluminum industry. The Steelworkers and the three major aluminum companies agreed to reduce normal retirement age from sixty-five to sixty-two and to provide cost-of-living supplements to retired workers, in addition to substantial increases in the level of benefits.[17] It is quite likely that this "breakthrough," as it has been termed by President Abel of the Steelworkers, will soon be adopted by many other industries.

Other kinds of issues seem to be emerging as subjects for collective bargaining. There has been some experimentation around the country with a reduced work week, generally in the form of longer workdays but fewer days in the week. The increased emphasis upon leisure, coupled with considerable unemployment, may bring renewed emphasis on shorter hours; it may also bring renewed pressure for increasing the overtime premium. Workloads and the pace of work have risen to the surface as issues in a number of instances, in a manner reminiscent of the textile industry in the days of the "stretchout." It is to be anticipated that efforts will be made in collective bargaining to circumscribe managerial prerogative in this area.

KEY BARGAINS

In assessing the impact of collective bargaining—at least in respect to wages and major fringes—upon the economy as a whole, major consideration must be given to the pattern-setting or standard-setting effects of key bargains. What happens in a settlement between a major employer in a major industry and the union representing its employees is likely to be reflected substantially—in terms of the overall package if not in detail—in other settlements in that industry, and in other industries as well. During World War II, in the course of implementing national wage-control policies, the National War Labor Board recognized intraplant and interplant inequities as proper bases for making wage adjustments. In the effort to secure approval for wage increases necessary to attract labor in a period of massive labor shortages, employers stressed the inequities from which some of their employees suffered vis-à-vis others and, even more, the inequities suffered by them vis-à-vis enterprises with higher wages. Unions, somewhat differently motivated, found it helpful to justify demands for higher wages upon similar claims of inequity. Absolute wage comparisons were often not persuasive with the regulatory authorities, tending to be offset by concepts of historic wage differentials. Much more to the point were comparisons of wage movements predicated upon the implicit notion that at some rather vague period in the past, the wage rates of Group A employees in Company B which was in Industry C stood in a fair and reasonable relationship to a comparable group of employees in a comparable enterprise in the same or in a comparable industry; that this relationship was altered by wage adjustments in the latter so that an inequitable condition resulted; that it was essential to restore the preexisting relationship by an appropriate adjustment for the now-disfavored or disadvantaged group.

This is, to be sure, an oversimplified and crude portrayal of the argument. It is not to be doubted, however, that in the exigencies of the times comparisons of this sort had a considerable currency—as a practical, rather than theoretical, matter—and a considerable influence upon both public policy and public opinion. This is not to say that comparisons of wage movements were devoid of economic significance or that they sprang full-blown from the tripartite National War Labor Board. But, given the climate of the period, they appealed widely as a fair and intelligent basis for wage setting. The appeal did not disappear when the war ended. On the contrary, despite a host of exceptions and variations, it has persisted to this day and comparisons of the sort described continue to be a major, though often unacknowledged, datum in collective bargaining. This can be seen in the fact that both public and private research agencies publish reports on the results of bargaining, in the recommendations or awards of fact-finding bodies or arbitrators, and in the bargaining postures of labor and management.

Important strategic and psychological, as well as economic, factors are involved. An employer who deals with a number of different unions (as in railroads, construction, and printing) is going to have a hard time persuading one union to settle for significantly less (absolutely or percentage-wise) than he has agreed to pay to other unions. And if he settles for more, he will almost surely have to go back and renegotiate the earlier settlements.

The instinct of workmanship has its place in collective bargaining. It will simply not do to bring to one's members a poorer settlement than has been attained by one's opposite number; one must do at least as well. In these terms, the test of the effectiveness of a negotiator lies less in the absolute gains he achieves than in the relationship between his "product" and that of the others with whom comparisons are customarily made. Similar motivations are present among management negotiators. A bargain which seems very advantageous when made may look prodigal to the point of stupidity if a relevant competitor strikes a better bargain.

Obviously, economic factors ("taking wages out of competition") are typically involved, and they are typically of vital significance. They are supplemented, however, by others. Thus, sanitation men in New York bargain with both eyes on what the policemen have won, and the wages of bus drivers in Los Angeles may be affected by what bus drivers are being paid in Chicago or Philadelphia. The UAW will seek a major gain from one or another of the major automobile companies with the expectation that, if it is successful, it will have little difficulty securing like gains from the others. Perhaps more significantly, what happens in automobiles, or steel, or electrical equipment is virtually certain to have far-reaching reflections in other industries, though it is impossible to forecast the precise ripple effects. Uniformity in result throughout the economy is not to be anticipated, but uniformity is not necessary for the generation, from two or three critical sources (industries or individual enterprises of vast size and importance to the economy), of influences—profound and immediate—over the whole economy.

Heightened emphasis upon "follow-the-leader" practices in wage negotiation is leading to interesting developments in bargaining unit structure, at least *de facto*. The awareness that the outcome of wage bargaining will be significantly affected, if not definitively determined, by some other bargain is leading to the adoption of a variety of plans ranging from the most informal kind of consultation to the formal enlargement of the bargaining unit. Where an employer negotiates with a number of unions within a single plant, there is a significant tendency for the unions to consult with and advise the union which is the first to go into the bargaining; they may be invited to sit in with the bargainer's policy committee and to attend bargaining sessions. It is to be assumed that the same thing occurs when a number of employers are involved in bargaining with a single union. It sometimes happens that bargaining goes on simultaneously by an employer with several unions or by a union with several employers. Here, again, the different bargains are interrelated, even if there is no economic compulsion to link them, with the growing likelihood of consultation and coordination of bargaining postures.

Informal action of this sort, in which the several parties retain their freedom of action and in which the bargaining units remain as distinct entities, is likely to be followed by more substantial linkages. Thus, in the 1972-73 strike on the Long Island Railroad, twelve of the "non-operating" unions formed themselves into a coalition, bargaining together under an agreement obligating them to unity of action. In the West Coast construction industry, a number of unions have for some years bargained as a group with the contractors. In the telephone industry, the Bell System has indicated that, in 1974, it plans to engage in nationwide bargaining for its affiliated companies with the various unions having representation rights in the affiliates.

Concerted action does not necessarily lead to merger of unions or to the establishment of employers' associations, for there are substantial obstacles to be overcome before permanent action of this sort can be taken. On the other hand, the advantages often appear to be considerable on each side of the bargaining table. In any event, bargaining units are becoming larger and are covering more establishments and more jobs.

The movement towards multiunion and multiestablishment bargaining with its obvious emphasis upon uniformity of terms of employment over a growing area necessarily carries with it a growing loss of flexibility of adaptation to local conditions. It is difficult, if not impossible, to draft a national agreement which would suit local preferences and usages on such matters as the particular holidays to be observed, the distribution of overtime, and the retention of seniority. Provision must be made, therefore, for such local practices, for otherwise the national agreement may give rise to more controversy than it settles. To provide the desired flexibility, a growing number of national agreements allow for variations on such matters to be worked out at the plant level through a "local supplement." Under a number of agreements, as in the automobile industry, agreement on such a "local supplement" is required before the national agreement becomes applicable to the individual plant.

ARBITRATION: GRIEVANCES AND CONTRACTS

The most impressive development in collective bargaining since World War II has been the use of arbitration for the settlement of controversies over the meaning and application of collective agreements—what is often referred to as grievance arbitration or the arbitration of rights. The submission of labor disputes for final and binding resolution, though often urged, had been little used in the United

States prior to World War II outside of a few well-organized industries like clothing, printing, and railroad transportation. Until an employer recognized the union or at least indicated some willingness to deal with it, there was no room for arbitration. The passage of the Wagner Act and the broad constitutional support given it by the Supreme Court, in compelling employers to recognize the majority representative as the exclusive representative of all the employees and in making it an unfair labor practice for the employer to refuse to bargain collectively, opened the door somewhat. There remained major legal impediments. Thus, in many sections of the country the old view still obtained, even after the end of World War II, that collective-bargaining agreements were nothing more than "gentlemen's agreements" which were unenforceabpf the courts. Also, many courts, reflecting the ancient hostility of law courts to arbitration, adopted the position that agreements to arbitrate future disputes were unenforceable. Before arbitration could be considerably extended, the legal problems had to be resolved: that is, collective-bargaining agreements viewed as valid and enforceable contracts and arbitration provisions made operative even where they were sought to be applied to future disputes or to matters not considered to be justifiable.

A great impetus was given to arbitration by the National War Labor Board during World War II. As an implementation of the nationwide no-strike policy, the NWLB routinely imposed upon the parties the obligation to submit disputes over interpretation and application to arbitration. By the end of the War, large numbers of employers, many of them formerly bitterly opposed to arbitration, had had some years of experience with it on a day-by-day basis and came to the conclusion that *grievance* arbitration was much to be preferred to *grievance* strikes. Actually, when President Truman called his labor-management conference in the fall of 1945, virtually the only item on which there was bilateral agreement was the wisdom of providing in collective-bargaining agreements for the arbitration of grievances. Fortunately, there was available a substantial number of arbitrators who had acquired invaluable experience with the War Labor Board, so that the end of the war did not compel the abandonment of the various arbitration plans which had been established during the war: the arbitrators were there, the lawyers and others concerned with labor-management relations on both sides were there, the need to avoid work stoppages was there, as well as the recognition by employers that unionism had come to stay as part of the American scene. It is quite likely that, in light of these circumstances, grievance arbitration would have found grm acceptance in American industry, even if there had been no other major influences at work. However, various legal developments were instrumental in giving to arbitration a dominant position in labor relations.

In 1947, Congress passed the Labor Management Relations Act, generally called the Taft-Hartley Act. Section 301(a) of that statute provided:

> Suits for violation of contracts between an employer and a labor organization representing employees in an industry affecting commerce as defined in this Act, or between any such labor organizations, may be brought in any district court of the United States having jurisdiction of the parties, without respect to the amount in controversy or without regard to the citizenship of the parties.

Section 301(a) and related provisions of the Act were authoritatively construed by the Supreme Court in the *Lincoln Mills* [18] case, which concerned the refusal of an employer to arbitrate a grievance under an agreement containing a no-strike clause. In holding that the collective-bargaining agreement and the agreement to arbitrate were enforceable, Mr. Justice Douglas wrote:

> There has been considerable litigation involving §301 and courts have construed it differently. There is one view that §301(a) merely gives federal district courts jurisdiction in controversies that involve labor organizations in industries affecting commerce, without regard to diversity of citizenship or the amount in controversy. Under that view §301(a) would not be the source of substantive law; it would neither supply federal law to resolve these controversies nor turn the federal judges to state law for answers to the questions. Other courts—the overwhelming number of them—hold that §301(a) is more than jurisdictional—that it authorizes federal

courts to fashion a body of federal law for the enforcement of these collective bargaining agreements and includes within that federal law specific performance of promises to arbitrate construction of §301(a) which means that the agreement to arbitrate grievance disputes, contained in this collective bargaining agreement, should be specifically enforced. . . .

Plainly the agreement to arbitrate grievance disputes is the *quid pro quo* for an agreement not to strike. Viewed in this light, the legislation does more than confer jurisdiction in the federal courts over labor organizations. It expresses a federal policy that federal courts should enforce these agreements on behalf of or against labor organizations and that industrial peace can be best obtained only in that way.

Lincoln Mills's great importance to labor law arises from the fact that the enforceability of collective-bargaining agreements and of undertakings to arbitrate was an essential threshold ruling for the developing law of arbitration. It was shortly followed by a number of other Supreme Court decisions [19] which, taken together, laid down the propositions that it is federal policy to promote arbitration as an instrument of industrial peace, that the agreement to arbitrate is the *quid pro quo* for a no-strike agreement, that an implicit no-strike agreement may be deduced from a provision for the settlement of grievances by arbitration, that the no-strike agreements may be enforced by injunctions, notwithstanding the Norris-LaGuardia Act, that the law governing arbitration is federal law though it may be administered by state, as well as by federal, courts, and that, as a general proposition, courts are to stay out of arbitration beyond determining whether or not there has been agreement to arbitrate the dispute in question, and even here doubts are to be resolved in favor of arbitration. A recent example of great potential significance is the decision of the Court in *Gateway Coal Company* v. *United Mine Workers* [20] in which it was held that disputes over mine safety were comprehended within the arbitration clause of the agreement, that the union was bound to arbitrate the issue, and that a strike by the miners was enjoinable, despite Norris-La Guardia.[21]

Heightened status for arbitration has also been forthcoming from the National Labor Relations Board. There are numerous grievances which may involve not only questions as to whether the collective-bargaining agreement has been violated but also possible violations of the Labor Act. In such instances, and assuming that the Board has the authority to refuse to entertain cases which have been heard (as potential contract violations) by an arbitrator or can be so heard, the question becomes one of Board policy. It had long been Board policy, under the so-called *Spielberg* [22] doctrine, to refuse to consider, *de novo,* cases which had been heard by an arbitrator and disposed of in a manner consistent with Board policy. In 1971, in *Collyer Insulated Wire,*[23] the Board went a good bit further, deciding that where a collective-bargaining agreement provided for arbitration it would defer consideration of the unfair labor practice, thus in effect compelling resort to the arbitration process.[24]

It has been estimated that about 95 percent of collective-bargaining agreements provide for the arbitration of grievances. Court concern with the process is limited to the determination of arbitrability: that is, whether the parties have entered into an agreement to arbitrate. Judicial review of arbitrators' awards is quite limited, and the awards are enforceable in the courts. Unions and employers, who are the parties in arbitration, have available a broad variety of arrangements; they may decide to use a single "permanent umpire," or a small panel from which arbitrators are chosen in rotation; they may select arbitrators from lists supplied by a designating agency; they may give the arbitrator broad discretion and they may exclude particular issues. Basically, it is *their* instrument for industrial peace, to be tailored to their needs and purposes.

Notwithstanding some criticism of grievance arbitration for alleged incompetence or dishonesty of arbitrators,[25] for growing formality and costs, and for delays, it is apparent that grievance arbitration has won nearly universal acceptance by labor and management. Matters stand quite otherwise with contract arbitration or the arbitration of interests, where the issue is what the terms of the agreement should be rather than what are the rights under an existing agreement. In some industries (e.g., street transportation and printing), it has not been uncommon for the parties to submit unresolved issues as to the terms of employment to arbitration. And in the period immediately after World War II, there was some state experimentation with laws compelling such arbitration in public utility industries as a means of averting major work stoppages.[26] However, the resistance to contract arbitration has been very great on the part of both labor and management, who have been most fearful of the risks involved in delegating to outsiders the authority to fix the basic terms of employment. Hence, while there has been extensive use of mediation, fact-finding, and impasse panels, contract arbitration has made no perceptible headway.[27]

This situation may now be in the process of change. The 1973 agreement between the steel industry and the United Steelworkers to submit to arbitration any terms of employment unresolved in their 1974 negotiations is of enormous potential significance. Although the authority of the as-yet-unnamed arbitrator has been defined with great particularity and his discretion materially circumscribed, this path-breaking venture does involve significant risks of an award which is unacceptable to one or the other party. On the other hand, an award which is satisfactory to both sides may well prove to be a stimulus to imitation in other segments of the economy, paving the way for the wider utilization of arbitration in contract disputes.[28]

UNION GOVERNMENT

Criticism of unions for undemocratic government and for corrupt practices became particularly acute in the years following World War II as part of a comprehensive attack upon unionism, which, it was alleged, had become too powerful a factor in American society. Perhaps the best-known of these criticisms was that of the McClellan Committee which, over a period of years, conducted an elaborate investigation into the so-called improper activities in the labor or management field.[29] There was not much in its reports which was either new or startling, and, on balance, the disclosures concerned a relatively small number of unions. Nevertheless, the groundwork was laid for legislation—the Labor-Management Reporting and Disclosure Act of 1959 (Landrum-Griffin) [30]—which declared that:

The Congress further finds, from recent investigations in the labor and management fields, that there have been a number of instances of breach of trust, corruption, disregard of the rights of individual employees, and other failures to observe high standards of responsibility and ethical conduct which require further and supplementary legislation that will afford necessary protection of the rights and interests of employees and the public generally as they relate to the activities of labor organizations, employers, labor relations consultants, and their officers and representatives. (Sec. 2[b].)

It is beyond the scope of this paper to discuss the several provisions of the statute, their justification, or the interpretations which the courts have placed upon them. Nevertheless, it is to be observed that large areas which formerly were wholly unregulated, or regulated solely by the states, have now been placed within the ambit of federal labor law. In the process, the potential membership participation has been immeasurably enlarged. With the guarantees of civil rights within the union, with the right to go to court against the union, and, perhaps most important, the regulation of elections and election procedures (with the potential for intervention by the Secretary of Labor), union administrations have become far more vulnerable to membership pressure and to challenge by the membership than ever before. Whether or not it is true, as union leaders often complain, that the statute has had the effect of hurting "good" unions and not touching "bad" unions, it seems fair to conclude that increased responsiveness to the membership and enlarged opportunities for rank-and-file militancy to replace "the establishment" are among the fruits of Landrum-Griffin.

In this respect, it may be conjectured that there will be little or no *sustained* increase in member participation. Under the conditions of contemporary large-scale unionism, the ideal of town-hall democracy is unattainable.[31] While members are, on the whole, undoubtedly loyal to their unions, their attitude toward day-in-day-out participation is notably one of apathy.[32] And, even if the attitude were otherwise, it is simply futile, given the size of the unions and the complexity of the functions they perform, to yearn for member participation on a broad and sustained scale.

POLITICAL ACTIVITY

It is a foregone conclusion that the labor movement will continue to expand its political activities at all levels of government, especially at the national level.[33] While there is not the faintest suggestion of the formation of an independent labor party, and while the traditional alliance between organized labor and the Democratic party may be subject to strains, there are major forces at work compelling increased participation by organized labor, directly or indirectly, in the political process. The large growth in union organization among public employees is one such force. Of broader significance is the vast range of governmental action which is of vital concern to unions and their members: wage-and-hour legislation, social security, occupational health and safety, labor relations laws, civil rights laws—to mention but a few. In such instances, there is a direct impingement of government action. In other areas, the relations are less direct, but there is a growing awareness within the labor movement of the significance of many matters it was formerly content to ignore. For instance, among the topics dealt with in the resolutions adopted on October 19, 1973, by the tenth annual AFL-CIO convention were: the federal tax and revenue system, the supply of money and credit, interest rates, foreign trade, export controls, energy, and public-investment programs.[34]

Moreover, there is a growing awareness of the likelihood of persisting governmental efforts at economic stabilization with their inevitable effects upon collective bargaining. As one scholar has said in discussing the problems inherent in balancing full employment, price stability, and collective bargaining:

> Because each of these three objectives is held firmly, though in some uncertain degree, by various groups within our society, we cannot dispense with any of them. What we will have is something of (but something less than) all of them. We will have not-quite-full employment, not-quite-stable prices, not-quite-free collective

bargaining. The degree of each will depend on the political bargains which are struck at a particular time. . . . Whether they like it or not, national union politicians will have to take more responsibility for wage-price controls (or, more broadly, an incomes policy) than they have in the past. Otherwise, they will forgo any constructive voice in determining how full employment should be, how much prices should be restrained, how discretionary collective bargaining may be.[35]

In the light of such pressures, the question is not whether labor will engage in politics but rather the form of such engagement and its size.[36] It is likely that the outstanding role here will be played by the AFL-CIO COPE (Committee on Political Education),[37] though individual unions may be expected to mount independent campaigns from time to time, especially in those instances in which there is a division of opinion within the movement as to the party and candidate to be supported.

CONCLUSION

The list of subjects dealt with above is not by any means exhaustive, omitting, for want of space, such matters as the changing role of strikes and race and sex discrimination.[38] It should be plain nevertheless that, on an overall view, American unionism has adhered faithfully to the philosophy of business unionism. It has reacted pragmatically to the world in which it lives and operates, emphasizing short-run considerations and eschewing doctrinaire positions. Its policies arise out of its "bread-and-butter" fundamentals, reflecting what is currently believed to be attainable. There is little ground for anticipating any material modification in the fundamental goals or the adoption of a broader fundamental outlook. But policies and tactics will keep on changing as new problems emerge and as the dimensions of old ones are altered by developments in our dynamic society.

NOTES

1. R. F. Hoxie, *Trade Unionism in the United States* (New York: Appleton, 1917).
2. Samuel Gompers, *The American Labor Movement* (Washington: A.F. of L., 1914), p. 21.
3. From time to time, "left-wing unionism"—for example, the I.W.W., the Trade Union Unity League and the Communist-dominated unions expelled from the CIO by Philip Murray—have appeared on the scene. They have seldom lasted long. More significantly, while they functioned, they stressed "business union" objectives even more forcefully than their opposite numbers in the AFL and tried to keep their political activities from public view so far as they could.
4. The Clayton Anti-Trust Act of 1914 was called the "Magna Carta" of American labor by Samuel Gompers, since it promised relief from criminal prosecutions, injunctions, and damage suits under the Sherman Act.
5. The change in the attitude of the Supreme Court began early in 1937, and was marked particularly by the decisions in the Wagner Act cases: see *N.L.R.B.* v. *Jones & Laughlin Steel Corp.*, 301 US 1 (1937).
6. As a part of the constitutional "revolution," the Court abandoned the concept of substantive due process under which much legislation had previously been invalidated on the ground that it was arbitrary, capricious, or unreasonable.
7. See, for example, Executive Order 11491, as amended by Executive Order 11616 of August 26, 1971.
8. See, R. C. Brown, "Public Sector Collective Bargaining: Perspective and Legislative Opportunities," *William and Mary Law Review,* Vol. 15, No. 1 (Fall 1973), pp. 57-92.
9. S. D. Spero and J. M. Capozzola, *The Urban Community and its Unionized Bureaucracies* (New York: Cambridge University Press, 1973), esp. chs. 10-12.
10. National Labor Relations Board, *37th Annual Report . . . 1972* (Washington, 1973), Chart 15 and Table 17.
11. *Ibid.,* p. 13.
12. *Ibid.,* p. 4.

13. See *Compliance, Enforcement and Reporting in 1972 under the LMRDA* (Washington: U.S. Department of Labor, Labor-Management Services Administration, 1973).
14. P. Henle, "Reverse Collective Bargaining? A Look at Some Union Concession Situations," *Industrial and Labor Relations Review*, Vol. 26 (1973), pp. 956-68. A striking illustration is afforded by the concessions made by the International Brotherhood of Teamsters, in March 1974, to induce a buyer to purchase the Rheingold Brewery in Brooklyn, the former owner having acted to shut the plant down permanently and to go out of the brewery business.
15. M. E. Segal Company *Newsletter*, Vol. 18, No. 1 (January 1974).
16. *Ibid.*
17. *The New York Times*, February 2, 1974.
18. *Textile Workers Union* v. *Lincoln Mills*, 353 US 448 (1957).
19. *United Steelworkers* v. *Warrior & Gulf Navigation Co.*, 363 US 574 (1960); *United Steelworkers* v. *American Mfg. Co.*, 363 US 564 (1960); *United Steelworkers* v. *Enterprise Wheel and Car Corp.*, 363 US 593 (1960); *Local 174* v. *Lucas Flour Co.*, 369 US 95 (1962); *Smith* v. *Evening News Assn.*, 371 US 195 (1962).
20. 38 L. Ed. 2d 583 (1974).
21. See also *Boys Market Inc.* v. *Retail Clerks*, 398 US 235 (1970).
22. 122 N.L.R.B. 1080 (1955).
23. 192 N.L.R.B. 150 (1971).
24. W. J. Isaacson and W. C. Zifchak, "Agency Deferral to Private Arbitration of Employment Disputes," *Columbia Law Review*, Vol. 73 (1973), pp. 1383-1418.
25. See Paul A. Hays, *Labor Arbitration: A Dissenting View* (New Haven: Yale University Press, 1966).
26. The Supreme Court invalidated these laws on the ground that their prohibition of strikes entailed an unconstitutional conflict with the provisions of the National Labor Relations Act. See *Amalgamated Assn. of Street Railway Employees* v. *WERB*, 340 U.S. 383 (1951).
27. See National Academy of Arbitrators, *Arbitration of Interest Disputes* (Washington: Bureau of National Affairs, 1974).
28. Negotiations began on January 30, 1974, to replace the agreement which expires August 1. Workers were guaranteed in any event a wage increase of at least 3 percent. *The New York Times*, January 31, 1974. An agreement was reached in April 1974. The parties agreed to follow the same procedure when the contract expires in 1977.

29. See *Sen. Rep. No. 1417,* 85th Cong. 2d Sess. (Washington: 1958), and *Sen. Rep. No. 621,* 86th Cong. 1st Sess. (Washington: 1959).

30. Public Law 86-257, 86th Cong. 1st Sess. (1959).

31. Emanuel Stein, "The Dilemma of Union Democracy," *The Annals,* Vol. 350 (November 1963), pp. 46-54.

32. G. W. Brooks, *The Sources of Vitality in the American Labor Movement* (Ithaca: New York State School of Industrial and Labor Relations, 1960).

33. H. H. Wellington, *Labor and the Legal Process* (New Haven: Yale University Press, 1968), ch. 6.

34. *The National Economy, 1973* (Washington: AFL-CIO, 1973).

35. N. W. Chamberlain, "Collective Bargaining in the Private Sector," *The Next Twenty-Five Years of Industrial Relations,* Gerald G. Somers, ed. (Madison, Wis.: Industrial Relations Research Association, 1973), pp. 23-24.

36. J. D. Greenstone, *Labor in American Politics* (New York: Knopf, 1969).

37. Phillip Taft, "Internal Union Structure and Functions," in *The Next Twenty-Five Years of Industrial Relations, op. cit.,* p. 9.

38. Discrimination because of race and sex is an ancient problem and is the subject of a great body of literature as well as of litigation. See, for example, *Steele* v. *Louisville & Nashville R.R. Co.,* 323 U.S. 192 (1944). Adjustments to current public policy on race and sex discrimination pose major problems to unions vis-à-vis "nonminority" members and vis-à-vis employers in bargaining on antidiscrimination provisions. They also present potential conflicts between unions and the agencies like the Equal Employment Opportunity Commission which are concerned with the elimination of forbidden discrimination under statutes like the Civil Rights Act of 1964, Title VII, and between antidiscrimination legislation and the National Labor Relations Board. See, for example, "Title VII and NLRA: Protection of Extra-Union Opposition to Employment Discrimination," *Michigan Law Review,* Vol. 72, No. 2 (December, 1973), pp. 313-31.

THREE

Is Industrial
Peace Achievable?

David L. Cole

Chairman
National Commission for
Industrial Peace

INTRODUCTION—THE CURRENT SETTING

As industrial history of the past thirty years has shown, shifts and developments in labor relations, particularly in the subjects encompassed in collective bargaining or in their changing emphasis, make it clear that this is a factor of growing importance and that it must be carefully taken into account in predicting or planning for the business future.

We have come through the past year or two surprisingly well in terms of labor strife and, on the whole, of the quality of the wage and economic settlements that have been made.

But there is a good deal of restlessness, to which the wage-price controls have contributed. There are sectors in which protests against discrimination are strongly expressed. Others are convinced they have been denied their opportunity and ability to achieve what their economic strength and the market forces would have permitted. The controls have now been watered down considerably. It is becoming easier to assert there have been signs of apparent discrimination and that the time has come to correct this and to catch up. (The controls were terminated on April 30, 1974—Ed. note.)

John Dunlop, with his usual resourcefulness, tried to salvage as much as possible in the declining stages of the cost-of-living control program. For our purposes it is significant to note that some decontrols were made conditionally. The price side is not our prime interest in this analysis. We are much more interested in the efforts to prevent industrial warfare and runaway labor costs on an industry-by-industry basis. The establishment of standing labor-management committees which will serve to moderate disputes and to lessen the danger of shutdowns in some of the more critical industries is one of the conditions imposed. The general threat to reimpose controls if things get out of hand is another.

Considering history, one would be justified in being gravely concerned about the course of industrial relations in a year in which the labor agreements in the following industries expire or are reopened: steel, aluminum, can, longshore, telephone, clothing, aerospace, coal, construction, airlines, and railroads, not to mention a host of local transit and other utilities.

In addition, we are facing demands that the emergency caused by the petroleum shortage entitles other groups to reopen their labor agreements, although they were not subject to reopening in 1974. The Teamster Union agreements are a conspicuous example.

THE NATURE OF INDUSTRIAL PEACE

Industrial peace is not measured by the incidence of strikes or other forms of work stoppages. The number of man-days lost in a stipulated period does not necessarily reflect the state of labor relations. The percentage of work time lost may be low, but many questions must be answered before we can say this means things are tranquil or in good order.

A number of questions must be answered. What kind of year was it, and in what areas were the strikes? Were the shutdowns in defense or other industries affecting national health or safety? Were they in basic industries which caused dislocations and unemployment in other industries? Were they essentially private quarrels or were they of the kind that had serious effects on the community or the economy? What was the atmosphere within industries and within the workplaces?

There may be a relatively strike-free period in a given enterprise or industry but a restless and discordant relationship nevertheless. Friction or a sense of hostility at the workplace causes inefficiency, interfering with the amount and quality of production and with the ability to maintain a relaxed relationship. There have been many demonstrations of this situation in our experience. For one thing, in such an atmosphere we find inordinate amounts of time devoted to grievances. The employees and their stewards swamp the grievance procedure by filing as many as possible. The supervisors react in an antagonistic way. They decline to make concessions, simply denying the grievances and, regardless of their merits, sending them to higher levels. The resulting delays increase the feelings of resentment. The employees retaliate by resisting management in a variety of ways. Innovations and new methods or technology are not given a willing trial. There are deliberate interferences with production or operations, some subtle and some quite open, through slowdowns or even wildcat walkouts. Schedules tend to be disrupted, and ill will enlarged.

We then have a snowballing effect on both sides. The workers become more impatient and less willing to listen to their union leadership. Efforts to restore good will become more difficult. Instead of good will percolating down to the people on the jobs, ill will is apt to rise from the workplace by a sort of capillary action to the leadership. This is paralleled on the side of management. The next contract negotiation is approached with distaste and animosity, and the outlook cannot be good.

Thus, although statistically the strike picture may not be bad, the condition is hardly one of peacefulness or one in which the prospect for cooperation or harmony can be called favorable. Industrial peace means far more than having a small percentage of man-days lost in a given period of time. The outlook, the atmosphere, and the attitudes of those engaged in or conducting the enterprise are of far greater importance. These are not easily measured, and certainly not by a simple mathematical computation.

Active strife in our current stage of labor relations is merely a reflection of what the conduct of most parties or their predecessors has been historically. It takes a good deal of determination and goodwill to break the tradition of suspicion and hostility that has prevailed. Recognition and status have been and to some degree are still the goals opposed to the conviction of many in management that it is their intrinsic right to determine what they should pay as the price for labor. Underlying is the clash of economic interests—how the fruits or revenues of the industry or enterprise should be divided between the employer and the employees. This lends itself readily to exploitation by appeals to our cruder self-interests.

It is still a matter of pride for one side to achieve what is regarded as a victory over the other. The result is that collective bargaining is often used as a strategic weapon rather than as a way of accommodating to each other's needs. The party which believes it has superior strength at the moment is quite willing to have a test of economic strength through a shutdown of operations. The union in such a case will often maintain that the employer is not bargaining in good faith and that there is no choice but to strike, or the employer may contend that the employees' right to strike should not be denied them, that it may have a therapeutic value and be a good thing. Such judgments, as stated, are based simply on feelings of superior strength. The union leadership may be convinced that the company's operations can be shut down, that its customers will demand relief, and that the economic pressures will give the employees a definite advantage. The company, on the other hand, may know that because of new technology it is possible to operate for an indefinite period with supervisory and technical forces, or it may feel that because of its multiplant structure or the financial condition of the union the advantage will be on its side.

Such a course reflects a reach for immediate victory with little thought of long-range consequences.

Settlements arrived at under such circumstances are the product of imposition rather than accord. They do little to reduce ill will, and feelings of resentment or injustice persist, with the weaker party considering itself taken advantage of and biding its time to correct things or to get its satisfaction. Hostilities of the kind prevalent in the earlier, more primitive, stages of the relationship are thus renewed or prolonged and stand in the way of any program of more rational or constructive labor relations or genuine industrial peace.

NATURE AND PURPOSE OF COLLECTIVE BARGAINING

Underlying a sound program of industrial peace should be an understanding of the nature and purpose of collective bargaining, and a willingness of leadership to conduct themselves accordingly.

The process of collective bargaining is now accepted and in fact declared to be the keystone of our industrial relations system. Rejected or strongly resisted by many segments of American industry only thirty or forty years ago, it is now universally embraced as the vital force in what we like to think of as our democratic industrial society.

Often it is referred to as free collective bargaining. In recent years with increasing frequency the qualification of responsibility has been added. While there are differences as to the meaning of "free and responsible" as applied to collective bargaining, it has generally been acknowledged that it should be responsive to the public or common interest.[1]

But confusion has crept into the views as to the very nature of collective bargaining. It seems clear that it is a process of reasoning and persuasion through which are composed the differences or conflicting interests of employers and employees that arise primarily when labor agreements are being developed or negotiated. Over 80 percent of the man-days lost are caused by such disputes. The persuasion in some situations is induced by the pressure of economic strife—the strike or lockout or kindred techniques—but there is general accord that reasoning is the basic ingredient of the kind of collective bargaining contemplated by our national labor policy. This characteristic is commonly acclaimed and sought to be preserved and strengthened as the essential feature of our process of collective bargaining.

These observations are made because there are those who speak of the strike as though this were the purpose or objective of collective bargaining. Of course, it is not. It is merely one of the means used to induce agreement, and there is surely no doubt that the responsible leadership of American labor and management, let alone the public, believe that it should be of diminishing importance and consequence as labor relationships mature and as the parties become more sophisticated and accustomed to dealing with one another.

Those who equate collective bargaining directly with the strike seem indifferent to the fact that they are championing a process that would be essentially punitive and destructive rather than rational and constructive.

The right to strike is quite different from the constant reliance on the strike. Constant reliance on the strike repudiates the reasoning and persuasion theory and is incompatible with the process of constructive collective bargaining. To suggest that we cannot have collective bargaining without reliance directly on the strike as the moving force is like saying that in international relations if we renounce warfare we cannot have diplomacy. In both situations precisely the contrary is true.

It cannot be emphasized too strongly that the fundamental purpose of collective bargaining is to find ways of reaching accord on the matters of concern to management and labor. It is an agreement-making process and as such is predicated on the willingness of both parties to agree. Such a process by its very nature must be essentially rational. It is by no means an immutable principle that we must rely on tests of economic strength. There are better ways of avoiding or composing differences in our complex and changing economy.

The obstacles are greatest where the power of either side is excessive, giving that side what it conceives to be an immediate advantage. It then tends to resist any form of restraint that might be imposed on its use of economic force. To many, this is the way the game is played.

THE PROGRESS TOWARD INDUSTRIAL PEACE

The above descriptions of labor relations and the current outlook may seem to present a bleak picture. It is not intended to suggest, however, that we are without hope.

Hardly a generation ago it was commonplace for industry and labor to resist each other as to almost every item that was proposed by either side, and it was more or less expected that every dispute might result in a strike, lockout, the discharge of employees, or possibly the discontinuance or removal of the plant.

With the passage of time the edge has been worn off many of the issues. It has gradually been acknowledged that it is not essential that they be resolved by shutting down operations. It will suffice to give a brief description of types of issues, formerly major causes of strikes, which are now customarily resolved or determined by other means.

By statute or other form of government regulation serious disputes are now settled by National Labor Relations Board decisions or elections in matters of union recognition, representation rights, and bargaining units or improper interferences in such matters. We now have legislation relating to minimum wages, hours, and overtime, unemployment compensation, old age retirement, prevailing wages for government work or work on government contracts, the retirement plan for railroad workers, and a variety of industrial safety and health measures. By agreement in almost every labor-management relationship, unresolved grievances are now concluded by binding arbitration instead of becoming the subject of strike. In the composite, this adds up to a most impressive transformation, and we should not consider ourselves at the end of this promising road by any means.

We must constantly bear in mind that the basic concept of collective bargaining in our type of society is that it is a process of trying to arrive at accord with a minimum or economic strife. As such, the parties must be willing to try to agree. Each retains the right to decline to agree. Voluntarism on both sides is the essential characteristic.

This point is emphasized because of the constantly repeated thought that strikes could be avoided by the enactment of compulsory arbitration laws.

Such laws are strongly opposed by both labor and management, with relatively few exceptions. The exceptions are largely in the area of public utilities and government employment. The opposition to compulsory arbitration, for a variety of reasons, is mainly that it would be incompatible with our system of free collective bargaining and industrial democracy, and that it ignores the fact that we have always proceeded and must still proceed by the process of agreement. Imposition by law runs counter to the very essence of the proposition that we must look to the parties to reach agreement.

Neither labor nor management is prepared to have some government agency dictate the terms and conditions of employment. The resistance to wage and price controls is an example of how distasteful permanent government dictation would be.

In 1972, emergency legislation as to labor disputes in the transportation industries was proposed and strongly supported by the Nixon administration. Its key features was the selection by a panel of the last offer of either the industry or the union. The opposition was substantially based on the ground that it amounted to compulsory arbitration and that this was unacceptable in whatever form in our system of free collective bargaining.

In my appearance before the Senate Committee I argued against its enactment, suggesting that legal compulsion is inconsistent with the freedom of choice by way of agreement, and that legal compulsion has not had a good record even in the field of public employment where the strike has by no means been traditional. Transportation strikes have caused great concern, and it has been necessary on a number of occasions for Congress to step in and set up *ad hoc* procedures to end railroad strikes or strike threats. However, since Congress would retain that right, we might better put our faith in voluntary activities or programs of management and labor, which currently give promise of being developed and which probably would be abandoned in protest if the proposed legislation were enacted.

A selection of one offer or the other by the panel, without giving it any discretion to modify, would lead to the selection of an entire labor agreement of immense complexity by individuals who would be denied the right to know what was discussed or proposed during negotiations.[2] This I consider to be extremely unwise. It would not be an agreement but something in the nature of a verdict or decision, and one based on evidence withheld from those making the decision. Such an agreement could well be intolerable to one side or the other, and it could not command respect. It would serve largely to set up a period of fermentation, leading to an assured blowup in the course of time.

Skeptics will say it is naïve to expect the parties to develop peaceful dispute-settling procedures. I believe, however, that in our economy this is our best choice and that both industrial history and current trends support the hope that improvement and relief can be achieved by this means.

Let us first understand that strikes cannot be completely eliminated, and recognize that many strikes can be endured without harm that is too grave, particularly those in the noncritical industries or operations. Even in a critical industry, a strike of moderate duration is not too heavy a price to pay for the preservation of our system of free determination of wages and conditions of employment.

Moreover, contrary to common belief, there have been a great many demonstrations over the years that labor and management are willing to forgo the strike or lockout and employ other means of settling or avoiding disputes over wages and other conditions of employment.

In the public utility field in at least two unions, the Amalgamated Street Railway and International Brotherhood of Electrical Workers, the locals were required by constitution or internal regulations to proffer arbitration before being permitted to go on strike. This remained in force for more than fifty years.

For some fifteen or twenty years after World War II ended, with no concerted or formal program, the voluntary submission of contract disputes to binding arbitration was not unusual. I served more than fifty times as arbitrator in such proceedings in a half-dozen industries.

Many of us have forgotten, moreover, that in England over a century ago the more important industries and their trade unions voluntarily instituted, and for more than twenty years adhered to, a program of voluntary arbitration of wage and other contract disputes.[3]

INTERUNION DISPUTES

There is another area that should be mentioned, although it represents an activity primarily within labor circles. This is in the matter of raiding and jurisdictional and work assignment disputes. We are only too aware of the innumerable and punishing strikes or other interferences with operations that have arisen out of such disputes over the years.

Not only are there provisions in the National Labor Relations Act relating to the regulation of work assignment disputes. Of perhaps greater consequence and significance are voluntary agreements. One such agreement entered into by the building trades affiliates of the AFL-CIO and major employer groups is that leading to the National Joint Board procedure. This has been in existence for twenty years and has resulted in cutting down both in number and impact the jurisdictional disputes and strikes with which this industry has been plagued for a long time.

The AFL-CIO Internal Disputes Plan is another major step of this kind. Originally, this was in the form of an agreement between the AFL and the CIO reached in 1954 on the eve of their merger. It is now part of the AFL-CIO constitution and is binding on all the affiliated unions and their locals and other subordinate bodies. It defines and deals with raids and attempts to take over work and provides for a ruling by a third party as the impartial umpire which is enforceable through specified sanctions.

One need not engage in involved research to be impressed with the significance of these voluntary steps by labor. Jurisdiction and the claims to given types of work have long been subjects jealously and bitterly fought over by unions, and the willingness to surrender a large part of such determinations to third parties represents a major move.

ALTERNATIVES TO ECONOMIC CONFLICT

Mention has been made above of the extensive use in the past twenty-five years of voluntary arbitration of wage and other terms of new or reopened agreements. Beyond this, a variety of other techniques have also been employed by labor and management in several industries to cut down the areas of difference and to assist in reaching agreement. Bok and Dunlop [4] list nine such procedures in their discussion of alternatives to economic conflict. As they point out, some of these are of recent origin, and some are practically as old as collective bargaining itself. These nine procedures are:

1. Prenegotiation conferences to shape proposals and negotiations (the garment, clothing and hosiery industries)
2. Joint selection of professional specialists to make cost estimates
3. Early negotiations well in advance of the deadline (Armour & Co.)
4. Private mediation including possible fact-finding and recommendations by jointly selected neutrals (Kaiser Steel Long-Range Plan)
5. National private joint machinery within an industry to mediate or settle terms of agreement (Industrial Relations Council of electrical branch of construction since 1919)
6. Continuing joint study committee for periodic discussions of issues (basic steel)
7. A formula-type arrangement for measuring prevailing wages (United Shoe Machinery Corp.)
8. An advance commitment by contract to submit unsettled issues to arbitration
9. Voluntary arbitration without advance commitment.

There have also been various combinations of these techniques, and innumerable contract issues have been resolved or avoided by their use.

Of perhaps greatest importance is the current acknowledgment by highly respected leaders of labor and management that in several basic or critical industries the strike is becoming an outmoded and undesirable means of resolving contract-making disputes. In arriving at this conclusion, George Meany emphasizes the excessive burden and cost of strikes in disputes between large and well-organized unions and industries and the fact that by submission of remaining issues to a mutually selected arbitrator the workers can achieve substantially what they could by resorting to a costly stoppage of work. He also speaks favorably of the necessity of unions and their members to be concerned about the welfare of the industry in their own self-interest in light of competition. He deplores the adverse effects of stockpiling caused by strike threats with the resulting loss of future business and job opportunities.

The outstanding case in which this reasoning has led to a concrete contractual program is of course that in the basic steel industry. For the first time since World War II, the industry and the United Steelworkers have assured each other and the users of their products that they need not be concerned over the danger of a strike in the 1974 negotiations. Although dissidents in the union attacked the Experimental Negotiating Agreement, a contract was agreed upon before the expiration date. In the aluminum and can industries, agreement was reached without a strike, and the wage adjustments and the innovations and improvements in fringe items helped to establish the outlines of the basic steel settlement.

Spokesmen for the steel companies and the steel union, particularly Heath Larry, vice-chairman of U.S. Steel, and I. W. Abel, president of the union, have gone to great lengths to explain the reasons for the course upon which they have embarked, and the central point is the need to protect the competitive position and future of the industry upon which they mutually depend.

Another industry in which the top labor leadership has publicly recognized the need to avoid shutdowns is the maritime industry, both on-shore and off-shore. Industry spokesmen are of course heartily in accord that American maritime commerce has suffered greatly over the years because of strikes, to the detriment of both the industry and the employees. Shipments have been tied up, vessels have been transferred to foreign-flag ownership, job opportunities have declined, and a huge volume of water-borne transportation has been lost.

Paul Hall, president of the Seafarers Union; Thomas Gleason, president of the International Longshoremen's Association; and Jesse Calhoon, president of the Marine Engineers' Association have joined with other labor representatives in proposing programs which they hope will permit them to reach settlements in their next negotiations without resort to strikes. They have declared that industry stability is essential and that they favor long-term contracts, with uniform expiration dates, that negotiations should be conducted without threat of strike, that interim wage reopenings should be of an automatic kind, and that they would give serious consideration to the establishment of a mechanism for the resolution of disputes without stoppages, presumably some form of third-party assistance.

Labor disputes in the maritime industry have been the subject of numerous Taft-Hartley emergency injunctions—more than those in any other industry. If the program just outlined is effectuated and made successful, it will be a historic accomplishment. Like the current effort in basic steel, it could serve as a pattern and could encourage others to consider a similar approach.

These are by no means the only programs of this kind under consideration. In the vital railroad industry, representatives of the two sides have been meeting well in advance of the next contract negotiation to discuss mutual problems and how to improve their cumbersome grievance procedure, and in other ways to anticipate and try to correct possible sources of disagreement or friction. To a greater or lesser degree, other industries are trying to follow a similar course.

In addition, we note more than a nominal amount of acceptance of the voluntary arbitration procedure in connection with wages or contract terms. There are standing provisions for this in some airline agreements. Currently the Federal Mediation and Conciliation Service routinely reports mediation efforts concluded by the parties agreeing to submit their remaining issues to arbitration. More conspicuous have been the accounts of the use of arbitration in salary disputes involving baseball players. Thirty potential big league holdout cases were avoided in February 1974 by means of arbitration.

THE 1974 LABOR PICTURE

In considering the outlook for 1974 we find some unfavorable factors but some others that are of a more optimistic character.

The cost-of-living controls program ended April 30th. Because of the great rise in the consumer price index in the recent past, it is expected that many unions will now press for catch-up increases. In many areas, the labor market was tight yet paradoxically because of the energy shortage earnings tended to go down as employment or hours of employment declined, or as supplies or transportation facilities were impaired. Some labor agreements have provisions for reopenings in case of emergency, and it is maintained that the energy problem constitutes such an emergency. Some labor organizations insisted that this unanticipated condition justified reopening even where there was no such contract provision. The relief promptly offered the independent truckers when they protested against the adverse effects of the fuel shortage and the regulations on this subject in the winter of 1974 are likely to encourage others to use self-help in the form of economic weapons for similar purposes.

It is because of the importance of this consideration that the President announced that he had designated William Usery, director of the Federal Mediation and Conciliation Service, as his special assistant to handle labor aspects of the energy problem.

On the other hand, the 1974 aluminum and can industry labor agreements were concluded without strikes. There are catch-up elements in the wage increases for the first year, but only moderate increases in the second and third years. This is partly due to reliance on the cost-of-living escalation provisions. The major new feature of both agreements is the liberalization of the retirement programs, especially in the cost-of-living adjustments for retired employees, and the acceleration of the retirement age to sixty-two with substantial increases in the amount of the monthly payments. Since the Steelworkers' Union played a leading part in reaching these agreements, they served as benchmarks in the basic steel negotiation that were successfully concluded without a strike. It is also likely that, as usual, steel will serve as a pattern setter for others.

Moreover, most of the agreements to be reexamined this year already have cost of living adjustment provisions, so that the catch-up feature will not be as serious as would seem likely at first blush.

There are, of course, some elements that are difficult to assess. One of these is the old problem of internal union politics, of the feeling that leadership must respond to the desires of the membership, particularly in these trying times, and to keep pace with other labor organizations which have done, or seem likely to do, unusually well this year.

In any event, in this uncertain year labor agreements in some of our most critical industries involving over 10 million workers will be up for adjustment of wages, among other things, 5.2 million in new contract negotiations and 4.9 million by way of deferred increases provided for in existing contracts.

CONCLUSIONS

Much as one might prefer it, there is no formula or miracle device for avoiding labor strife. Our system of labor relations is one in which the two sides must come to agreement, and obviously each is free to disagree. This does not mean, however, that economic strife is desirable or essential. As indicated above, those most deeply involved in collective bargaining, the authoritative and respected spokesmen of labor and industry, have been most actively and creatively searching for substitutes for the strike as the means of composing their differences.

We should first understand and accept the fact that in our type of society the strike is not about to be abolished. We should confine our concern to strikes in the basic industries or critical services where the adverse impact is strongly on the community or on the economy as a whole.

There is an array of techniques that have been employed over the years in one industry or area or another to minimize the use of strikes.

These procedures have no magic. One or the other has worked in one situation but not in another, and for a time, and then been abandoned. Each one can be pointed to as having a record of success somewhere or at some time.

It is not, however, the form of the procedure that can spell success. The form is definitely of minor importance. What is of prime importance is the genuine desire of the parties to arrive at agreement. If they have this desire and if they convince each other that they do not have an ulterior purpose, then any of a variety of procedures will suffice. The parties can then readily fashion or adapt the procedures to meet their needs.

It is most constructive for the parties to do this; it clearly represents a valid exercise of the functions of collective bargaining. Such an approach is not for altruistic reasons. It is usually dictated by self-interest. But once the principals are able to persuade themselves that they no longer have to continue to fight the old war in the old way, then it becomes quickly evident that self-interest and mutual interest overlap and frequently to a large extent coalesce. The steel and maritime industries are good examples.

But we must be realistic and not overlook the lessons of history and experience. For well-established, traditional reasons leadership which advocates a nonviolent approach is suspect in many quarters. It takes more courage to agree than to disagree, regardless of the merits.

We must be careful not to overlook the primary and vital step in the transition that is being advocated. If the parties must approach the problems in labor relations with a disposition or willingness to try to come to voluntary accord, their task must not be made harder, or indeed practically impossible, by legislation or other legal steps of a compulsory nature. Such steps are inflammatory, and the reaction can easily be anticipated.

This is not to suggest that certain legal restrictions have not worked. Those were, however, of a kind more or less favored by both industry and labor, or at least not seriously opposed, and amount to what has been called compulsion by acquiescence. Examples of matters in which the strike has practically been eliminated in favor of elections or other forms of decisions have been described above. But the critical point which should be underlined is that in this field it is not possible to create goodwill by government fiat, nor to eliminate ill will by legislative mandate.

I summarized above nine dispute-settling procedures. None of these forms in itself is satisfying, as I review my experience. The missing ingredient is the disposition or desire of the parties to have the procedure succeed in bringing them to agreement.

My strong preference is for an indefinite combination of all these procedures, together with any others the parties may devise that they believe may be helpful. I would favor a course in which they move by agreement step by step from one procedure to another until their differences are ended. I would exclude nothing, nor would I compel them to go forward if either chooses not to do so. This is because of the very nature of agreement-making.

The use of third parties in whatever capacities or functions the parties desire should not be excluded. But before moving into the next step, the neutral should make sure both parties want him to take it. The possible procedures are well known.

It was essentially this approach which was used in a number of airline-pilot contract-making disputes. It worked well in every one of some half-dozen cases until it was wrecked because of a bitter jurisdictional feud between the pilots and the flight engineers.

Because its popularity seems to be rising again, the arbitration of contract-making disputes deserves special mention. Arbitration is conducted by choice of both parties but in which, nevertheless, the award is legally binding. This device has been in extensive use for periods of time, and then for intervals of varying duration has been more or less in discard. Periodically, interest seems to revive, as occurred in the 1950s; we now seem to be at the beginning of another cycle.

This procedure has been criticized. It has been argued that it is too informal in character, that there are no rules of evidence or procedure, and that there are no criteria or standards to guide the arbitrator in arriving at his decision. Such arbitrations were criticized in England during the nineteenth century largely on the basis that the lack of guiding standards made the awards subjective and dependent on the personal predilections of the arbitrator.

The critics make the mistake of judging such arbitration as a form of litigation and by comparing it with judicial procedures. In fact, no law or existing contract is interpreted or applied in these proceedings, there is no decision of right or wrong, and consequently it lacks the basic quality of litigation.

What we have is a situation in which two parties who must arrive at agreement are unable to do so. They have the alternative of trying to impose a point of view on one another by force of economic power or by calling in some third party, in whose capacity and integrity they have confidence, to resolve their differences. He does not make a decision in the conventional sense. Rather, he determines what in his judgment the parties should have agreed upon in light of all the pertinent facts and circumstances. In other words, he acts as their jointly designated agent to make for them the agreement which at the time they are unable to make by themselves.

This is how I have always regarded my function in such arbitrations, and it has seemed to me that this is what the parties want.

As to the point that such awards are unpredictable, my response is to ask just how predictable can an agreement be which is exacted or imposed by one party on the other only if its strike, lockout, or kindred technique prevails.

I have no desire to magnify the importance of the neutral. It is still the parties themselves who will create the necessary atmosphere, indicate the proper disposition, and either make or break the negotiations. The third party really serves mainly as a peacemaker. He induces the parties to reason with one another with a minimum of rancor. His recommendations, when he is asked to make any, either in procedure or in substance, come to him mainly from the parties as he observes and hears them. As was said almost a century ago by Sidney and Beatrice Webb, it is not in the wisdom of his decisions that an arbitrator in such disputes makes his contribution, but in his ability to be an effective conciliator.

The concept that the vital force in a constructive labor relations program is the mutual desire of the parties to compose their differences, with procedures distinctly secondary, is borne out by the inconclusiveness of our national emergency labor legislation which has remained unchanged for the past twenty-seven years. The preventive mediation program of the Federal Mediation and Conciliation Service and the program of the labor-management advisory committee of the American Arbitration Association are also based primarily on this thought.

There are many who would like to see strong governmental action in this field. As indicated above reliance must be placed on the parties, with appeals to their self-interest, because at this stage there is more hope in this approach than in the police type of approach.

Only passing reference is made in this analysis to the factor of social responsibility. This is not because there is no such responsibility. It has been rather latent, but it can be aroused, and it probably will be. In former years, oddly, the sense of social responsibility played an important part in a great many labor disputes. Unions, and managements as well, were very conscious of the public's interest in public utilities and government services, as well as in other vital industries, and certainly so in times of national need. This factor has tended to be subordinated in recent years to other considerations, but the need to give it due weight has been stressed over and over. Reference was made earlier to the report of President Kennedy's Advisory Committee on Labor-Management Policy. The same point was repeatedly made in a 1961 report of a distinguished study group set up by the Committee for Economic Development.[5]

It may not be inappropriate to quote a prediction made in 1956 by Walter Reuther:

> the goal that we need to shoot for, hoping to achieve it before 1975, but making that the target date. Labor and management must recognize that in a free society they may discharge their broad social and moral responsibility only as they jointly succeed in elevating collective bargaining above the level of a struggle between competing economic pressure groups, and recognizing that basic decisions in collective bargaining must be based upon the total needs of the total community. . . .

> ". . . [W]e need to recognize that in a free society, bargaining decisions should be based upon facts and not upon economic power. I hope the day will come in America when, in collective bargaining problems and other problems that bear upon the

economic interest, decisions can be based upon the power of economic persuasion rather than upon the persuasion of economic power. In the exercise of naked economic power we make arbitrary decisions, which too frequently are in conflict with the basic needs of the whole of our society.[6]

My conclusions are based on our current and most recent experiences and studies. Circumstances and time may call for modifications. As William H. Davis, one of the wisest of all mediators, once observed: "In labor relations the tools wear out or the people just get tired of them and from time to time we have to find new ones."

NOTES

1. *Free and Responsible Collective Bargaining and Industrial Peace:* Report to the President from Advisory Committee on Labor-Management Policy (Washington: Government Printing Office, May 1, 1962).
2. *National Emergency Disputes, 1971-72* (Washington: Senate Subcommittee on Labor, March 28, 1972), p. 939.
3. Sidney and Beatrice Webb, *The History of Trade Unionism,* Reprint of Economic Classics (New York: Augustus M. Kelley, 1965), p. 337.
4. Derek C. Bok and John T. Dunlop, *Labor and the American Community* (New York: Simon and Schuster, 1970), p. 242.
5. *The Public Interest in National Labor Policy,* A Report by an Independent Study Group, Released by Committee for Economic Development (New Jersey, 1961).
6. Walter Reuther, *Labor's Role in 1975, U.S. Industrial Relations: The Next Twenty Years* (Lansing, Mich.: Michigan State University Press, 1956), p. 63.

FOUR

Business
and Technology

Peter F. Drucker

Marie Rankin Clarke Professor of Social Science
Claremont Graduate School
Claremont, California

Technology has been front-page news for well over a century—and never more so than today. But for all the talk about technology, not much effort has been made to understand it or to study it, let alone to manage it. Economists, historians, and sociologists all stress the importance of technology—but then they tend to treat it with "benign neglect," if not with outright contempt. (On this, see the "note" at the end of this paper.)

More surprisingly, business and businessmen have done amazingly little to understand technology and even less to manage it. Modern business is, to a very considerable extent, the creature of technology. Certainly the large business organization is primarily the business response to technological development. Modern industry was born when the new technology of power generation—primarily water power

99

at first—forced manufacturing activities out of home and workshop and under the one roof of the modern "factory," beginning with the textile industry in eighteenth-century Britain. And the large business enterprise of today has its roots in the first "big business," the large railroad of the mid-nineteenth century, that is, in technological innovation. Since then, the "growth industries," down to computer and pharmaceutical companies of today, have largely been the outgrowth of new technology.

At the same time, business has increasingly become the creator of technology. Increasingly, technological innovation comes out of the industrial laboratory and is being made effective through and in business enterprise. Increasingly, technology depends on business enterprise to become "innovation"—that is, effective action in economy and society.

Yet business managers, or at least a very sizable majority of them, still look upon technology as something inherently "unpredictable." Organizationally and managerially, technological activity still tends to be separated from the main work of the business and organized as a discrete and quite different "R & D" activity which, while in the business, is not really of the business. And until recently business managers, as a rule, did not see themselves as the guardians of technology and as concerned at all with its impact and consequence.[1]

That this is no longer adequate should be clear to every business manager. It is indeed the thesis of this paper that business managers have to learn that technology is managerial opportunity and managerial responsibility. This means specifically:

1. Technology is not more mysterious or "unpredictable" than developments in the economy or society. It is capable of rational anticipation and demands rational anticipation. Business managers have to understand the dynamics of technology. At the very least, they have to understand where technological change is likely to have major economic impact and how to convert technological change into economic results.

2. Technology is not separate from the business and cannot be managed as such if it is to be managed at all. Whatever role "R & D" departments or research laboratories play, the entire business has to be

organized as an Innovative Organization and has to be capable of technological (but also of social and economic) innovation and change. This requires major changes in structure, in policies, and in attitude.

3. The business manager needs to be concerned as much with the impacts and consequences of technology on the individual, society, and economy as with any other impacts and consequences of his actions. This is not talking "social responsibility"—that is, responsibility for what goes on in society (e.g., minority problems). This is responsibility for impact of one's own actions. And one is always responsible for one's impact.

These last ten years there has been a widely reported "disenchantment with technology." It is by no means the first one in recent history (indeed, similar "disenchantments" have occurred regularly every fifty years or so since the mid-eighteenth century). What is certain, however, is that technology will be more important in the last third of this century and will, in addition, change more than in the decades just past. Such great needs as the energy crisis, the environmental crisis, and the problems of modern urban society make this absolutely certain. Indeed, one can anticipate, with high probability, that the next twenty-five years will see as much, and as rapid, technological change as in the "heroic age" of invention, the sixty years between the mid-nineteenth century and the outbreak of World War I. In that period, which began in 1856 with Perkins's discovery of aniline dyes and Siemens's design of the first workable dynamo, and which ended in 1911 with the invention of the vacuum tube and with it of modern electronics, today's "modern"—and even tomorrow's "postmodern"— worlds were born. In this "heroic age" a new major invention appeared on the average every fifteen to eighteen months, to be followed almost immediately by the emergence of a new industry based on it. The next twenty-five or thirty years, in all likelihood, will far more resemble this late-nineteenth-century period than the fifty years since the end of World War I, which, technologically speaking, were years of refinement and modification rather than of invention. To the business and the businessmen who persist in the traditional attitude toward technology, the attitude which sees in it something "mysterious," something "outside," and something for which other people are responsible, technology will therefore be a deadly threat. But to business and businessmen who accept that technology is *their* tool, but also their responsibility, technology will be a major opportunity.

ANTICIPATING AND PLANNING TECHNOLOGY

The "unpredictability" of technology is an old slogan. Indeed, it underlies to a considerable extent the widespread "fear of technology." But it is not even true that *invention* is incapable of being anticipated and planned. Indeed, what made the "great inventors" of the nineteenth century—Edison, Siemens, or the Wright brothers—"great" was precisely that they knew how to anticipate technology, to define what was needed and would be likely to have real impact, and to plan technological activity for the specific breakthrough that would have the greatest technological impact—and, as a result, the greatest economic impact.

It is even more true in respect to "innovation" that we can anticipate and plan; indeed with respect to "innovation," we have to anticipate and plan to have any effect. And it is, of course, with "innovation" rather than with "invention" that the businessman is concerned. Innovation is not a technical, but a social and economic, term. It is a change in the wealth-producing capacity of resources through new ways of doing things. It is not identical with "invention," although it will often follow from it. It is the impact on economic capacity, the capacity to produce and to utilize resources, with which "innovation" is concerned. And this is the area in which business is engaged.

It should be said that technology is no more "predictable" than anything else. In fact, predictions of technology are, at best, useless and are likely to be totally misleading. Jules Verne, the French science fiction writer of a hundred years ago, is remembered today because his predictions have turned out to be amazingly prophetic. What is forgotten is that Jules Verne was only one of several hundred science fiction writers of the late nineteenth century—which indeed was far more the age of science fiction writing than even the present decade. And the other 299 science fiction writers of the time, whose popularity often rivaled and sometimes exceeded that of Jules Verne, were all completely wrong. More important, however, no one could have done anything at the time with Jules Verne's predictions. For most of them, the scientific foundation needed to created the predicted technology did not exist at the time and were not coming into being for many years ahead.

For the businessman—but also for the economist or politician—what matters is not "prediction," but the capacity to act. And this cannot be based on "prediction."

But technology can be anticipated. It is not too difficult—though not easy—to analyze existing businesses, existing industries, existing economies and markets to find where a change in technology is needed and is likely to prove economically effective. It is somewhat less easy, though still well within human limitations, to think through the areas in which there exists high potential for new and effective technology.[2]

We can say flatly first that wherever an industry enjoys high and rising demand, without being able to show corresponding profitability, there is need for major technological change and opportunity for it. Such an industry can be assumed, almost axiomatically, to have inadequate, uneconomic, or plainly inappropriate technology. Examples of such industries would be the steel industry in the developed countries since World War II or the paper industry. These are industries in which fairly minor changes in process, that is, fairly minor changes in technology, can be expected to produce major changes in the economics of the industry. Therefore, these are the industries which can become "technology prone." The process either is economically deficient or it is technically deficient—and sometimes both.

We can similarly find "vulnerabilities" and "restraints" which provide opportunities for new technology in the economics of a business and in market and market structure. The questions "What are the demands of customer and market which the present technology and the present business and the present technology do not adequately satisfy?" and "What are the unsatisfied demands of customer and market?," that is, the basic questions underlying market planning, are also the basic questions to define what technologies are needed, appropriate, and likely to produce economic results with minimum effort.

A particularly fruitful way to identify areas in which technological innovations might be both accessible and highly productive is to ask: "What are we afraid of in this business and in this industry? What are the things which all assert 'can never happen,' but which we nonetheless know perfectly well might happen and could then threaten us? Where, in other words, do we ourselves know at the bottom of our hearts that our products, our technology, our whole approach to the satisfaction we provide to market and customer, is not truly appropriate and no longer completely serves its function?" The typical response of a business to these questions is to deny that they have validity. It is the responsibility of the manager who wants to manage technology for the benefit of his business and of his society to overcome this almost reflexive response and to force himself and his business to take these questions seriously. What is needed is not always new technology. It might equally be a shift to new markets or to new distributive channels. But unless the question is asked, technological opportunities will be missed, will indeed be misconceived as "threats."

These approaches, of which only the barest sketch can be given in this paper, apply just as well to needs of the society as to needs of the market. It is, after all, the function of the businessman to convert need, whether of individual consumer or of the community, into opportunities for business. It is for the identification and satisfaction of that need that business and businessmen get paid. Today's major problems, whether of the city, of the environment, or energy, are similar opportunities for new technology and for converting existing technology into effective economic action. At the same time, businessmen in managing technology also have to start out from the needs of their own business for new products, new processes, new services to replace what is rapidly becoming old and obsolescent, that is, to replace today. To identify technological needs and technological opportunities one also starts out, therefore, with the assumption that whatever their business is doing today is likely to be obsolete fairly soon.[3]

This approach assumes a limited and fairly short life for whatever present products, present processes, and present technologies are being applied. It then establishes a "gap," that is, the sales volume which products and processes not yet in existence will have to fill in two, five, or ten years. It thus identifies the scope and extent of technological effort needed. But it also establishes what kind of effort is needed. For it determines why present products and processes are likely to become obsolescent, and it establishes the specifications for their replacement.

Finally, to be able to anticipate technology, to identify what is needed and what is possible, and above all what is likely to be productive technology, the business manager needs to understand the dynamics of technology. It is simply not true that technology is "mysterious." It follows fairly regular and fairly predictable trends. It is not, as is often said, "science." It is not even the "application of science." But it does begin with new knowledge which is then, in a fairly well-understood process, converted into effective—that is, economically productive—application.

THE PACE OF TECHNOLOGY

It is often asserted today that technology is moving at a lightning pace, as compared with earlier times. There is no evidence for this assertion. It is equally asserted that new knowledge is being converted much faster into new technology than at any earlier time. This is demonstrably untrue. In fact, there is a good deal of evidence that it takes longer today to convert new knowledge, and especially new scientific knowledge, than it did in the nineteenth century—if not, indeed, in the eighteenth and earlier centuries. There is a lead time, and it is fairly long.

It took some twenty-odd years from Siemens's design of the first effective dynamo to Edison's development of the electric light bulb, which first made possible an electrical industry. It has taken at least as long, in fact it has taken longer, from the design of the first working computer in the early forties to the establishment of truly producing computers—let alone to the development of the "software" without which a computer is (as was the early electric company) a "cost center" rather than a producer of wealth and economic assets. And there are countless similar examples. The lead time for the conversion of new knowledge into effective technology varies greatly between industries. It is perhaps shortest in the pharmaceutical industry. But even there, it is closer to ten years than to ten months. And, in any one industry, the lead time seems to be fairly constant.

What has shortened is the time between the introduction of new technology into the market and its general acceptance. There is less time to establish a pioneering position, let alone a leadership position. But even there, the "lead time" has not shortened as dramatically as most people, including most businessmen, assume. For both the electric light bulb and the telephone, that is, for the 1880s, the lead time between a successful technological invention and widespread—indeed, world-wide—acceptance was a few months. Within five years after Edison had shown his light bulb to the invited journalists, every one of the major electrical manufacturing companies in existence today in the Western world (excepting only Phillips in Holland) was established, in business, and leaders in its respective markets. And the same held true for the telephone and for telephone equipment.

In other words, it is the job of the businessman to understand what new knowledge is becoming acceptable and available, to assess its possible technological impact, and to go to work on converting it into technology—that is, into processes and products. He has to know, for this, not only the science and technology of his own field. Above all, he has to know that major technological "breakthroughs" very often, if not usually, originate in a field of science or knowledge that is different from that in which the old technology had its knowledge foundations. In this sense, the typical approach to "research," that is, the approach for developing specialized expertise in the field in which one already is active, is likely to be a bar to technological leadership rather than its main pillar, as is commonly believed. What is needed, at least as a complement, is the ability to scan the horizon of knowledge, to be alert to new insights and new awareness, and to be able to see their potential application to one's own field. What is needed, in other words, is a "technologist," rather than a "scientist." And often a layman, with good "feel" for science and technology, and with genuine intellectual interest, does this much better than the highly trained specialist in a technical or a scientific field—who is likely to become the prisoner of his own advanced knowledge.

It is not necessary, it is indeed not even desirable, for the businessman to become a "scientist" or even a "technologist." His role is to be the manager of technology. This requires an understanding of the process of technology and of its dynamics. It requires willingness to anticipate tomorrow's technology and, above all, willingness to accept that today's technology with its processes and products is becoming obsolete rapidly. It requires identifying the needs for new technology and the opportunities for it, in the vulnerabilities and restraints of the business, in the needs of the market and in the needs of the society. Above all, it requires acceptance of the fact that technology has to be considered a major business opportunity, the identification and exploitation of which is what the businessman is paid for.

THE INNOVATIVE ORGANIZATION

The next quarter century, as has already been said, is likely to require innovation and technological change as great as any we have ever witnessed. Most of this, however, in sharp contrast to the nineteenth century, will have to be done in and by established organizations, and especially in and by established businesses.

It is not true, as is often said, that "big business monopolizes innovation." On the contrary, the last twenty-five years have been preeminently years in which small business and often new and totally unknown businesses produced a very large share of the most effective innovations. Xerox was nothing but a small paper merchant as late as 1950. Even IBM was still a small company and a mere pygmy, even in its own office equipment industry, as late as World War II. Most of today's pharmaceutical giants were either small companies at the end of World War II or barely in existence, and so on.

But still, increasingly, the major effort in technological change is development and the effort of market introduction. These do not require "genius." They require highly educated people in a massive cooperative effort. And they require very large sums of capital. And these are indeed found in established institutions, whether business or government.

Altogether the existing businesses will have to become innovative organizations. For the last fifty to seventy-five years our emphasis has, properly, been on managing what we already know and understand. For the pace of technological innovation—and even the pace of economic change—in these last seventy-five years was, contrary to popular belief, singularly slow.[4]

Now business will again have to become entrepreneurial. And the entrepreneurial function, as the greatest of Continental European economists, J. B. Say (1767-1832), saw clearly almost two centuries ago, is to move existing resources from areas of lesser productivity to areas of greater productivity. It is to create wealth not by discovering new continents, but by discovering new and better uses for the existing resources and for the known and already exploited economic potentials. And technology, while not the only tool for this purpose, is an important one and may well be the most important one.

The great task of business can be defined as counteracting the specific "law of entropy" of any economic system: the law of the diminishing productivity of capital. It was on this "law" that Karl Marx based his prediction of the imminent demise of the "capitalist system." Yet capital has not only not become less productive, it has steadily increased its productivity in the developed countries—contrary to the assumed "law." But Karl Marx was right in his premise. Left to its own devices, any economy will indeed move toward steadily diminishing productivity of capital. The only way to prevent it from becoming entropic, the only way to prevent it from degenerating into sterile rigidity, is the constant renewal of the productivity of capital through entrepreneurship—that is, through moving resources from less productive into more productive employment. This, therefore, makes technology the more important the more highly developed technologically a society and economy become.

In the next twenty-five years, when the world will have to grapple with a population problem, an energy problem, a resources problem, and a problem of the basic community, that is, the city, this function is likely to become increasingly more critical—independent, by the way, of the political, social, or economic structure in a developed economy, that is, independent of whether the "system" is "capitalist," "socialist," "communist," or something else.

This will require businessmen to learn how to build and how to manage an innovative organization.[5] Normally, the innovative organization is being discussed in terms of "creativity" and of "attitude." What it requires, however, are policies, practices, and structure. It requires, first, that management anticipate technological needs, identify them, plan for them, and work on satisfying them.

It requires, second, and perhaps more important, that management systematically abandon yesterday.

"Creativity" is largely an excuse for doing nothing. The problem in most organizations which are incapable of innovation and self-renewal is that they cannot slough off the old, the outworn, the no longer productive. On the contrary, they tend to allocate to it their best resources, especially of good people. And any body incapable of eliminating waste products poisons itself eventually. What is needed to make an innovative organization is a systematic policy for abandoning the no longer truly productive, the no longer truly contributing. The innovative organization requires, above all, that every product, every process, every activity, be put on "trial for its life" periodically—maybe every two or three years. The question should be asked: "If we did not do this already, would we now—knowing what we now know—go into it?" And if the answer is No, then one does not ask: "Should we abandon it?" Then one asks: "How can we abandon it, and how fast?"

The organization, whether business, university, or government agency, which systematically sloughs off yesterday need not worry about "creativity." It will have such a healthy appetite for the new that the main task of management will be to select from among the large number of good ideas for the new the ones with the highest potential of contribution and the highest potential of successful completion.

But beyond this, the innovative organization needs specific policies. It needs measurement and information systems which are appropriate to the economic reality of innovation—and a regular, moderate, and continuous "rate of return on investment" is the wrong measurement. Innovation, by definition, is only cost for many years, before it produces a "profit." It is first an investment—and a return only much later. But that also means that the rate of return must be far larger than the highest "rate of return" for which managers plan in a managerial type of business. Precisely because the lead time is long and the failure rate high, a successful innovation in an innovative organization must aim at creating a new business with its potential for creating wealth, rather than a nice and pleasant addition to what we already have and what we already do.

Finally, we will have to realize that innovative work is not capable of being organized and done within managerial components, that is, units concerned primarily with work on today and on tomorrow morning. It needs to be organized separately, with different structural principles and in different structural components. Above all, the demands on managerial self-discipline and on clarity of direction and objectives are much greater in innovative work and have to be extended to a much larger circle of people. And therefore, the innovative organization, while organically a part of the ongoing business, needs to be structurally and managerially separate. Businessmen, to be able to build and lead innovative organizations, will, therefore, have to be able to do both—manage what is already known and create the new and unknown. They will have to be able to optimize the existing business and to maximize the potential business.

These, to most businessmen, are strange and indeed somewhat frightening ideas. But there are plenty of truly innovative businesses around—in practically every country—to show that the task can be done, and is indeed eminently doable. In fact, what is needed primarily is recognition—lacking so far in most management thinking and in almost all management literature—that the innovative organization is a distinct and different organization, and is not only a slightly modified managerial organization.

RESPONSIBILITY FOR THE IMPACT OF TECHNOLOGY

Everybody is responsible for the impact of his actions. This is one of the oldest principles of the law. It applies to technology as well. There is a great deal of talk today about "social responsibility." But surely the first point is not responsibility for what society is doing, but responsibility for what one is doing oneself. And therefore, technology has to be considered under the aspect of the businessman's responsibility for the social impacts of his acts. In particular, there is the question of the "by-product impacts," that is, the impacts which are not part of the specific function of a process or product but are, necessarily or not, occurring without intention, without adding to the intended or wanted contribution, and indeed as an additional cost—for every by-product which is not converted into a "salable product" is, in effect, a waste and therefore a cost.

The topic of the responsibility of business for its social impact is a very big one.[6] And the impacts of technology, no matter how widely publicized today, are among the lesser impacts. But they can be substantial. Therefore, the businessman has to think through what his responsibilities are and how he can discharge them.

TECHNOLOGY ASSESSMENT

There is, these days, great interest in "technology assessment," that is, in anticipating impact and side effects of new technology *before* going ahead with it. The U.S. Congress has actually set up an Office of Technology Assessment. This new agency is expected to be able to predict what *new technologies* are likely to become important, and what long-range effects they are likely to have. It is then expected to advise government what new technologies to encourage and what new technologies to discourage, if not to forbid altogether.

This attempt can only end in fiasco. "Technology assessment" of this kind is likely to lead to the encouragement of the wrong technologies and the discouragement of the technologies we need. For *future* impacts of *new* technology are almost always beyond anybody's imagination.

DDT is an example. It was synthesized during World War II to protect American soldiers against disease-carrying insects, especially in the tropics. Some of the scientists then envisaged the use of the new chemical to protect civilian populations as well. But not one of the many men who worked on DDT thought of applying the new pesticide to control insect pests infecting crops, forests, or livestock. If DDT had been restricted to the use for which it was developed, that is, to the protection of human beings, it would never have become an environmental hazard; use for this purpose accounted for no more than 5 or 10 percent of the total at DDT's peak, in the mid-sixties. Farmers and foresters, without much help from the scientists, saw that what killed lice on men would also kill lice on plants, and made DDT into a massive assault on the environment.

Another example is the "population explosion" in the developing countries. DDT and other pesticides were a factor in it. So were the new antibiotics. Yet the two were developed quite independently of each other; and no one "assessing" either technology could have foreseen their convergence—indeed, no one did. But more important as causative factors in the sharp drop in infant mortality, which set off the "population explosion," were two very old "technologies" to which no one paid any attention. One was the elementary public health measure of keeping latrine and well apart—known to the Macedonians before Alexander the Great. The other one was the wire-mesh screen for doors and windows invented by an unknown American around 1860. Both were suddenly adopted even by backward tropical villages after World War II. Together they were probably the main "causes" of the "population explosion."

At the same time, the "technology impacts" which the "experts" predict almost never occur. One example is the "private flying boom," which the experts predicted during and shortly after World War II. The private plane, owner-piloted, would become as common, we were told, as the Model T automobile had become after World War I. Indeed, "experts" among city planners, engineers, and architects advised New York City not to go ahead with the second tube of the Lincoln Tunnel, or with the second deck on the George Washington Bridge, and instead to build a number of small airports along the west bank of the Hudson River. It would have taken fairly elementary mathematics to disprove this particular "technology assessment"—there just is not enough airspace for commuter traffic by air. But this did not occur to any of the "experts"; no one then realized how finite airspace is. At the same time, almost no "experts" foresaw the expansion of commercial air traffic at the time the jet plane was first developed or that it would lead to mass transportation by air, with as many people crossing the Atlantic in one jumbo jet twelve times a day as used to go once a week in a big passenger liner. To be sure, transatlantic travel was expected to grow fast—but of course it would go by ship. These were the years in which all the governments along the North Atlantic heavily subsidized the building of new superluxury liners, just when the passengers deserted the liner and switched to the new jet plane.

A few years later, we were told by everybody that "automation" would have tremendous economic and social impacts—it has had practically none. The computer offers an even odder story. In the late forties nobody predicted that the computer would be used by business and governments. While the computer was a "major scientific revolution," everybody "knew" that its main use would be in science and warfare. As a result, the most extensive market research study undertaken at that time reached the conclusion that the world computer market would, at most, be able to absorb 1000 computers by the year 2000. Now, only twenty-five years later, there are some 150,000 computers installed in the world, most of them doing the most mundane bookkeeping work.

Then a few years later, when it became apparent that business was buying computers for payroll or billing, the "experts" predicted that the computer would displace "middle management," so that there would be nobody left between the chief executive officer and the foreman. "Is middle management obsolete?" asked a widely quoted *Harvard Business Review* article in the early fifties; and it answered this rhetorical question with a resounding Yes. At exactly that moment, the tremendous expansion of middle-management jobs began. In every developed country middle-management jobs, in business as well as in government, have grown about three times as fast as total employment in the last twenty years; and their growth has been parallel to the growth of computer usage.

Anyone depending on "technology assessment" in the early fifties would have abolished the graduate business schools as likely to produce graduates who could not possibly find jobs. Fortunately, the young people did not listen and flocked in record numbers to the graduate business schools so as to get the good jobs which the computer helped to create.

But while no one foresaw the computer impact on middle-management jobs, every "expert" predicted a tremendous computer impact on business strategy, business policy, planning, and top management—on none of which the computer has, however, had the slightest impact at all. At the same time, no one predicted the real "revolution" in business policy and strategy in business in the fifties and sixties, the merger wave and the "conglomerates."

DIFFICULTY OF PREDICTION

It is not only that man no more has the gift of prophecy in respect to technology than in respect to anything else. The impacts of technology are actually more difficult to predict than most other developments. In the first place, as the example of the "population explosion" shows, social and economic impacts are almost always the result of the convergence of a substantial number of factors, not all of them technological. And each of these factors has its own origin, its own development, its own dynamics, and its own experts. The "expert" in one field, e.g., the expert on epidemiology, never thinks of plant pests. The expert on antibiotics is concerned with the treatment of disease—whereas the actual explosion of the birth rate largely resulted from elementary and long-known public health measures.

But, equally important, what technology is likely to become important and have an impact, and what technology either will fizzle out— like the "flying Model T"—or will have minimal social or economic impacts—like automation—is impossible to predict. And which technology will have social impacts and which will remain just technology is even harder to predict. The most successful prophet of technology, Jules Verne, predicted a great deal of twentieth-century technology a hundred years ago (though few scientists or technologists of that time took him seriously). But he anticipated absolutely no social or economic impacts, but an unchanged mid-Victorian society and economy. Economic and social prophets, in turn, have the most dismal record as predictors of technology.

The one and only effect an "Office of Technology Assessment" is likely to have, therefore, would be to guarantee full employment to a lot of fifth-rate science fiction writers.

THE NEED FOR TECHNOLOGY MONITORING

However, the major danger is that the delusion that we can foresee the "impacts" of new technology will lead us to slight the really important task. For technology does have impacts, and serious ones, beneficial as well as detrimental ones. These do not require prophecy. They require careful monitoring of the actual impact of a technology once it has become effective. In 1948 practically no one correctly saw the impacts of the computer. Five and six years later, one could and did know. Then one could say: "Whatever the technological impact, *socially* and *economically* this is not a major threat." In 1943 no one could predict the impact of DDT. Ten years later, DDT had become a worldwide tool of farmer, forester, and livestock breeder, and, as such, a major ecological factor. Then, thinking as to what action to take should have begun, work should have been started on the development of pesticides without the major environmental impact of DDT, and the difficult "trade-offs" should have been faced between food production and environmental damage—which neither the unlimited use nor the present complete ban on DDT sufficiently considers.

"Technology monitoring" is a serious, an important, indeed a vital task. But it is not "prophecy." The only thing possible, in respect to *new* technology, is *speculation* with about one chance in a hundred of being right—and a much better chance of doing harm by encouraging the wrong, or discouraging the most beneficial new technology. What needs to be watched is "developing" technology, that is, technology which has already had substantial impacts, enough to be judged, to be measured, to be evaluated.

And "monitoring" a "developing" technology for its social impacts is, above all, a managerial responsibility.

But what should be done once such an impact has been identified?

Ideally, it should be eliminated. Ideally, the fewer the impacts, the fewer "costs" are being incurred, whether actual business costs, externalities, or social costs. Ideally, therefore, businesses start out with the commitment to convert the elimination of such an impact into "business opportunity."

And where this can be done, the problem disappears, or rather it becomes a profitable business and the kind of contribution for which business and businessmen are properly being paid. But where this is not possible, business should have learned, as a result of the last twenty years, that it is the task of business to think through what kind of regulation is appropriate. Sooner or later, the impact becomes unbearable. It does no good to be told by one's public relations people that the "public" does not worry about the impact, that it would, in fact, react negatively toward any attempt to come to grips with it. Sooner or later, there is then a "scandal." The business which has not worked on anticipating the problem and on finding the right solution, that is, the right regulation, will then find itself both stigmatized and penalized—and properly so.

This is not the popular thing to say. The popular thing is to assert that the problems are obvious. They are not. In fact, anyone who would have asked for regulation to cut down on air pollution from electric power plants twenty or even ten years ago would have been attacked as an "enemy of the consumer" and as someone who, "in the name of profit," wanted to make electricity more expensive. (Indeed, this was the attitude of regulatory commissions when the problem was brought to their attention by quite a few power companies.) When the Ford Motor Company in the early fifties introduced seat belts, it almost lost the market. And the pharmaceutical companies were soundly trounced by the medical profession every time they timidly pointed out that the new high-potency drugs required somewhat more knowledge of pharmacology, biology, and biochemistry than most practicing physicians could be expected to have at their disposal.[7]

But these examples also, I think, bring out that the "public relations" attitude is totally inappropriate and, in fact, self-defeating. They bring out that neglect of the impacts and willingness to accept that "nobody is worried about it" in the not-so-very-long run penalizes business far more seriously than willingness to be unpopular could possibly have done.

Therefore, in technology-monitoring the businessman not only has to organize an "early-warning" system to identify impacts, and especially unintended and unforeseen impacts. He then has to go to work to eliminate such impacts. The best way, to repeat, is to make the elimination of these impacts into an opportunity for profitable business. But if this cannot be done, then it is the better part of wisdom to think through the necessary public regulation and to start early the education of public, government, and also of one's own competitors and colleagues in the business community. Otherwise the penalty will be very high—and the technology we need to tackle the central problems of "postindustrial" society will meet with growing resistance.

CONCLUSION

Technology is certainly no longer the "Cinderella" of management, which it has been for so long. But it is still to be decided whether it will become the beautiful and beloved bride of the "prince," or instead turn into the fairy tale's "wicked stepmother." Which way it will go will depend very largely on the business executive and his ability and willingness to manage technology. But which way it will go will also very largely determine which way business will go. For we need new technology, both major "breakthroughs" and the technologically minor but economically important and productive changes to which the headlines rarely pay attention. If business cannot provide them, business will be replaced as a central institution—and will deserve to be replaced. Managing technology is no longer a separate and subsidiary activity that can be left to the "longhairs" in "R & D." It is a central management task.

A HISTORICAL NOTE

The absence of any serious concern and study of technology among the major academic disciplines is indeed puzzling. In fact, it is so puzzling as to deserve some documentation.

The nineteenth-century economist usually stressed the central importance of technology. But he did not go beyond paying his elaborate respects to technology. In his system he relegated technology to the shadowy limbo of "external influences," somewhat like earthquakes, locusts, or wind and weather, and as such incomprehensible, unpredictable, and somehow not quite respectable. Technology could be used to explain away phenomena which did not fit the economist's theoretical model. But it could not be used as part of the model. The twentieth-century Keynesian economist does not even make the formal bow to technology which his nineteenth-century predecessor regarded as appropriate. He simply disregards it. There are, of course, exceptions. Joseph Schumpeter, the great Austro-American economist, in his first and best-known work on the dynamics of economic development, put the "innovator" into the center of his economic system. And the innor in large part was a technological innovator. But Schumpeter found few successors. Among living economists only Kenneth Boulding at the University of Colorado seems to pay any attention to technology. The ruling schools, whether Keynesian, neo-Keynesian, or Friedmanite, pay as little attention to technology as the preindustrial schools of economists, such as the Mercantilists before Adam Smith. But they have far less excuse for this neglect of technology.

Historians, by and large, have paid even less attention to technology than economists. Technology was more or less considered as not worth the attention of a "humanist." Even economic historians have given very little attention to technology until fairly recently. Interest in technology as a subject of study for the historian did not begin until Lewis Mumford's book *Technics and Civilizations* (New York: Harcourt, Brace, 1934). It was not until twenty-five years later that systematic work on the study of the history of technology began, with publication in England in 1957/58 of *A History of Technology,* edited by Charles Singer (London: Oxford University Press, 1954-58), five vols.; and simultaneously in the United States with the founding of the Society for the History of Technology in 1958 and of its journal, *Technology and Culture.* The relationship between technology and history has further been discussed in the first American textbook, *Technology in Western Civilization,* edited by Melvin Kranzberg and Carroll W. Pursell, Jr. (New York: Oxford University Press, 1967), 2 vols., and in my essay volume: *Technology, Management and Society* (New York: Harper & Row, 1970), (especially in the essays, "Work and Tools" in *Technology and Culture* (Winter 1959), "The Technological Revolution" and "Notes on the Relationship of Technology, Science and Culture" in *Technology and Culture* (Fall 1961) and "The First Technological Revolution and its Lessons," delivered as Presidential Address to the Society for the History of Technology in December 1965, and published in *Technology and Culture* (Spring 1966). The California medievalist, Lynn White, Jr., has done pioneering work on the impact of technological changes on society and economy, especially in his book, *Medieval Technology and Social Change* (London: Oxford University Press, 1962). But the only work that tries successfully to integrate technology into history, especially economic history, is the recent book by the Harvard economic historian David S. Landes, *The Unbound Prometheus: Technological Change and Industrial Development in Western Europe; 1750 to the Present* (London: Cambridge University Press, 1969). Outside of the English-speaking countries only one historian of rank has given any attention to technology, the German Franz Schnabel in his *Deutsche Geschichte im Neunzehnten Jahrhundert* (Freiburg: Lerder, 1929-37).

Perhaps even more perplexing is the attitude of the sociologist. While the word "technology" goes back to the seventeenth century, it first became a widely used term as slogan if not as manifesto of the early sociologists in the late eighteenth century. To call the first technical university in 1794 Ecole Polytechnique was, for instance, a clear declaration of basic principles and, above all, a declaration of the central importance of technology to society and social structure. And the early fathers of sociology, especially the great French sociologists, Saint-Simon and Auguste Comte, did indeed see technology as the great liberating force in society. Marx still echoes some of this—but then relegates technology to the realm of secondary phenomena. Sociologists since then have tended to follow Marx and to put the emphasis on property relationships, kinship relationships, and on everything else, but not on technology. There are plenty of slogans such as that of "alienation." But there has been practically no work done. And technology is barely mentioned in the major sociological theories of the last, that is, the post-Marx, century from Max Weber to Marcuse and Levy-Bruhl to Levi-Strauss and Talcott Parsons. Technology either does not exist at all for the sociologist, or it is an unspecified "villain."

In other words, the scholars have yet to start work on technology, as the way man works; as the extension of the limited physical equipment of the biological creature that is man; as a part—a major part—of man's intellectual history and intellectual achievement; and as a human achievement which, in turn, influences the human condition profoundly. However, the businessman cannot wait for the scholars. He has to manage technology now.

NOTES

1. The one exception to this has been Japan since the Meiji Restoration of 1867. From the early days, the Japanese saw technology clearly as part of economy and society and as something that deserved and required careful planning and deliberate management.
2. See my book, *Managing for Results* (New York: Harper & Row, 1964), esp. chs. 10 and 11.
3. The best discussion of the planning approach from the needs of the business is to be found in Michael J. Kami, "Business Planning as Business Opportunity," in *Preparing Tomorrow's Business Leaders Today,* ed. by Peter F. Drucker (Englewood Cliffs, N.J.: Prentice Hall, 1969). This essay, written for the symposium that celebrated the fiftieth anniversary of the Graduate School of Business Administration of New York University, sums up the experience of the two American companies which have most successfully innovated in technology in the period since World

War II. In the fifties and sixties, Mr. Kami served first as head of Long-Range Planning for IBM and then as head of Long-Range Planning for Xerox.

4. See my book, *The Age of Discontinuity* (New York: Harper & Row, 1969), esp. ch. 1 or 2.
5. See my book, *Management: Tasks; Responsibilities; Practices* (New York: Harper & Row, 1974), esp. ch. 61.
6. *Ibid.*
7. For more examples of the complexity of the impact problem, see the section, "Three Cautionary Tales," in *ibid.*, ch. 24.

FIVE

Productivity
and Business

John W. Kendrick

Professor of Economics
The George Washington University

The high, and rising, planes of living in the United States have largely been due to cost-reducing innovations by business firms and the resulting productivity increases in the business sector, which accounts for 85 percent of the GNP, as measured.

In this paper I shall review productivity trends in the United States private domestic economy since 1948, with particular attention to the period since 1966, when there appears to have been some tendency toward a slowing in rates of productivity advance. Then I shall summarize the basic factors behind increasing productivity, suggesting some of the causes of the slowdown in recent years. Finally, I shall offer several policy recommendations for improving the productivity performance of business.

PRODUCTIVITY TRENDS SINCE 1948

First let us look at productivity trends between 1948 and 1969—both cycle peak years with comparably low unemployment rates, averaging around 3.5 percent of the civilian labor force. By 1948 the postwar readjustment had largely been completed, the final vestiges of national price controls having been removed in 1947. And 1969 was the next to the last year before our recent experiments with wage and price controls, and the attendant distortions which followed.

Over the twenty-one-year period beginning 1948 total factor productivity—the ratio of output (real product) to labor, capital, and natural resource inputs, measured in real terms—increased at an average annual rate of 2.3 percent in the private domestic business economy. (See Table 5-1.) As I documented in my recent study,[1] 2.3 percent was approximately the same annual rate that had prevailed since World War I, after abstracting the effects of the Great Depression. It was about double the rates that prevailed before World War I.

Output per man-hour, the more usual "partial productivity" ratio, rose at the average rate of 3.1 percent of year from 1948 to 1969. This was a somewhat faster pace of advance than had prevailed earlier, due to a significantly higher rate of substitution of capital for labor after 1948. The faster growth of capital goods per man-hour reflects the more stable rate of growth after World War II than before.

Productivity changes vary over the business cycle, of course, even in the mild cycles we have had since 1948. Rates of change usually accelerate in the last stage of contractions and rise until capacity rates of utilization are reached in major industries. Productivity increases at a slower rate in the latter phase of expansion and through the early stages of contraction. The impact of these cyclical movements of productivity on unit labor costs and their relationship to prices, is an important part of the explanation of business cycles, as first pointed out by Wesley C. Mitchell over sixty years ago,[2] and recently documented for the period since 1948 by Geoffrey Moore.[3]

Rates of productivity change also vary from one cycle to another, whether measured from peak to peak or between cycle averages. In part, these variations reflect differential rates of economic growth generally, as well as unique characteristics of each subperiod. Thus, productivity advanced strongly in the 1948-53 period, and slowed down in the next two subperiods ending with 1960, as the battle against inflation slowed economic growth after 1955. Productivity then advanced strongly again in the 1960-66 period but showed the weakest performance of all from 1966 to 1969.

The productivity trend rate for the business economy is, of course, in effect a weighted average of rates of productivity change in all of the component industries. Rates of gain have varied widely among industries. At the high end of the scale (in terms of real product per man-hour) at between 5 and 6 percent a year are two manufacturing groups—petroleum refining and chemicals—and three nonmanufacturing industries, farming, telephone communications, and electric and gas utilities. Even higher are two other regulated industries—airlines, and pipelines—at 8 and 9 percent, respectively.

At the low end of the scale are the broad-finance and private-services groups, contract construction, and lumber and products, with between 1 and 2 percent a year. Leather products show near zero productivity gains, and local transit registered a decline both in output and in productivity. The other broad 2-digit groups came somewhere in between most within ± one percentage point of the 3.1 percent mean. When the industry groups are subdivided more finely, there is naturally greater dispersion of rates of change. The same is true of subperiods within the longer 1948-69 epoch.

The useful thing about having productivity estimates by industry is that it provides a handle for empirical analysis of the variables believed to be associated with productivity change. So in addition to time-series analysis one can use cross sectional data relating industry differences in rates of change in productivity to industry differences in levels or in rates of change of causal factors. We shall refer to the results of both types of analysis in the discussion of sources of productivity improvement.

The table also shows the rates of productivity change between 1969 and 1973. The year 1973 precedes a drop in production which may not be classified as a cyclical contraction by the National Bureau of Eco-

Table 5-1

PRODUCTIVITY TRENDS IN THE U.S. PRIVATE DOMESTIC ECONOMY
BY MAJOR INDUSTRY DIVISIONS
AVERAGE ANNUAL PERCENTAGE RATES OF CHANGE,
1948–1973, BY SUBPERIODS

	Period 1948–1969	Subperiods [1]					
		1948–1953	1953–1957	1957–1960	1960–1966	1966–1969	1969–1973p
Private domestic economy							
Real product	3.9	4.6	2.5	2.7	5.2	3.4	3.8
Total factor productivity	2.3	2.7	1.9	2.2	2.8	1.1	2.1
Real product per unit of capital	0.2	0.2	−1.1	0.2	1.7	−0.9	0.2
Real product per man-hour	3.1	4.1	2.7	2.6	3.6	1.7	2.9
Industry divisions: (Real product per man-hour)							
Agriculture	5.7	6.4	4.1	5.9	5.8	6.7	
Mining	3.8	5.2	3.3	4.7	3.7	1.8	
Contract Construction	1.3	4.4	3.2	1.5	−0.5	−3.0	
Manufacturing	3.0	3.7	2.2	2.2	3.6	2.7	
Durable goods	2.8	3.6	1.4	1.8	3.8	2.2	
Nondurable goods	3.3	3.5	3.4	2.9	3.4	3.4	
Transportation	3.2	2.2	3.1	3.0	4.8	2.2	
Communications	5.4	5.4	3.6	7.6	5.7	4.6	
Electric and gas utilities	5.9	7.6	6.3	5.4	5.1	4.4	
Trade	2.6	2.5	2.7	1.9	3.9	2.1	
Wholesale	3.1	2.2	3.3	2.9	3.7	3.0	
Retail	2.5	2.6	2.2	1.2	3.8	1.0	
Finance, Insurance & Real Estate	1.9	1.5	2.7	1.5	2.6	−0.4	
Services	1.1	0.5	1.2	1.1	1.7	0.4	

p = preliminary.
1. Subperiods are measured between successive business cycle peaks.

Source: John W. Kendrick, *Postwar Productivity Trends in the United States* (New York: National Bureau of Economic Research, 1973); estimates extended from 1969 through 1973 by the author.

nomic Research. But I believe that the subperiod 1969-73 yields useful information concerning basic trends.

Despite the fact that the rate of economic growth over the four-year subperiod was approximately the same as that over the preceding twenty-one years, the rate of productivity advance was about 10 percent lower—2.1 percent a year average for total factor productivity and 2.9 percent for real product per man-hour. Further, after the first quarter of 1973, output per man-hour in the private economy remained on a virtual plateau for the succeeding three quarters, sagging slightly in the fourth quarter as real GNP growth decelerated. Then with the sharp drop in real GNP in the first quarter of 1974, productivity also declined significantly. Even if real GNP grows for the rest of 1974 and scores a modest gain over 1973, it is doubtful if productivity will rise much, if any, above the 1973 mark.

Taking the sharp deceleration in productivity advance 1966-69 and the incomplete recovery 1969-73 together, it is clear that there was a significant deceleration in productivity advance over the seven-year period as a whole. The slowdown was greater than in the seven-year period 1953-60—1.6 versus 2.0 percent a year on average for total factor productivity, and 2.4 versus 2.7 percent for output per man-hour, even though the rate of economic growth was lower in the 1950s.

I am fully aware that the numbers may be interpreted differently, depending on how one combines subperiods. Nevertheless, my judgment is that there has been a significant deceleration in productivity advance since the mid-1960s. This does not mean that the trend rate in the future will continue lower than that of the past half-century. Future experience depends importantly on policies that will be adopted in coming years. By "running scared" and adopting policies favorable to productivity growth on the hypothesis that growth has been unsatisfactory, there will be a better chance for a favorable outcome than may be the case if we are complacent.

Before leaving the numbers it should be noted that output per man-hour in manufacturing since 1965 has increased much less in the United States than in eleven other advanced countries for which data are available. Between 1965 and 1970 the average annual rate of increase in the United States was 2.0 percent as compared with 6.7 percent for the eleven other countries. The differential was much less after 1970 and wage-rate increases exceeded productivity by a wider margin abroad than in the United States, so that our trends in unit labor costs and prices were favorable.[4] Nevertheless, the international comparisons provide another cause for concern over the U.S. productivity record.

CAUSAL FORCES BEHIND PRODUCTIVITY CHANGE

What are the causal forces behind productivity advance and what are the possible causes of the slowdown since 1966? The analytical framework for this discussion consists of four major headings:

1. Values and attitudes;
2. Institutional forms, policies, and practices;
3. Investments, tangible and intangible, designed to enhance the efficiency of the productive factors; and
4. Noninvestment forces.

Values and Attitudes

The basic value system of a society, which conditions its institutions and practices, obviously has an important bearing on productivity. The values and attitudes of the American people historically have been favorable to productivity advance. Most Americans have desired rising planes of living for themselves and for their children; they have been willing to work hard, save, invest, and take risks for material and social progress; and have generally sought to adapt to the social changes that technological progress entails. They have generally achieved the degree of social stability consistent with flexibility and diversity necessary to encourage long-term private investments, including investment in self, which raise productivity.

Since the early 1960s, however, there has been an increasingly critical attitude toward economic growth and "materialism." Possibly there has been some weakening of the work ethic and an increasing hedonism, associated with a spreading use of drugs and attendant increases in crime rates. There was a noticeable increase in social unrest associated initially with the civil rights movement and then with the anti-Vietnam war movement. Social divisiveness was, of course, promoted by the New and Old Left movements with their ultraliberal allies.

These negative social tendencies probably habxe bearing on the productivity slowdown after 1966, although the lack of adequate social indicators makes the impact difficult to measure. Also, despite the attention given to social tensions. I believe that the large majority of Americans have held to much the same values as in the past. Certainly, most Americans continue to strive for rising real incomes, although there is more awareness of the importance of the quality of life as well as the quantity of goods and services. Further, with the deescalation of the Vietnam conflict, by the early 1970s many of the social tensions were abating.

A more persistent, long-term problem relates to attitudes of workers toward increasing productive efficiency. Despite the drive for higher real incomes, in particular work situations employees—particularly organized labor frequently resist technological innovations or other management efforts to raise productivity if these are seen to threaten job security and/or offer little direct benefit to the affected workers. Our cross-sectional regression analyses for manufacturing industries show that, other things equal, rates of productivity advance have a significant *negative* correlation with degree of unionization. This is not surprising in view of the prevalence of restrictive union work rules, and efforts to control the pace of innovation. Further, attitudinal surveys show that most workers believe that productivity gains generally redound to the benefit of other groups, particularly stockholders. These findings pose a distinct challenge to managements to find ways to elicit greater cooperation from labor in raising productivity.

Institutions

The United States and many other countries rely on a predominantly market-directed competitive enterprise system based on private ownership and the profit motive. Theoretically, and I believe in practice, this system is a powerful engine of progress. Firms have the profit incentive to expand or to protect markets by introducing new and better products and cost-reducing ways and means of production. Widened profit margins tend to be competed away, of course, by other firms that imitate the innovators. This competitive pressure brings down unit real costs and product prices relative to factor prices, thus spreading the benefits of productivity advance. Under competition, the only way a firm can continue to enjoy abnormally high profit margins is continually to stay ahead of the pack by its progress technologically. Such profits are socially accepted as an incentive and reward for innovation.

Even if competition is more or less imperfect, as long as there is a significant degree of cross-elasticity of demand within product families and interindustry competition for the customer's dollar, the market system tends to reward technologically progressive firms and to penalize the inefficient. This view is supported by our regression analyses, which show no significant degree of correlation between industry productivity rates and the degree of concentration by industry, although concentration ratios are imperfect measures of the degree of competition.

Interestingly, the regulated industries show a generally higher rate of productivity gain than the nonregulated industries. In part, this may be due to greater use of capital and a greater potential for technological advance than for most other industries. In part, it reflects competition among the producers of capital goods for the electric utilities, transportation, and other regulated industries. But it also reflects the profit motive—in that the regulatory lag puts a premium on offsetting factor price increases by raising factor productivity in order to prevent or slow an erosion of allowable profit margins while waiting for rate adjustments by regulatory bodies.

The assumption by the federal government of responsibility for maintenance of high-level employment in the Employment Act of 1946 and the subsequent mitigation of business fluctuations has been a positive contribution to productivity advance. Less investment has been lost in contractions than in earlier periods, so that the growth of capital stocks has been stronger and steadier.

On the other hand, the very success of macroeconomic policy in maintaining relatively high employment has built-in an inflation bias, with its dangerous tendency toward acceleration which we witnessed

1966-70 and again since early 1973 despite direct wage and price controls. Now, efforts to counteract the unfavorable effects of inflation divert resources from other uses, and so tend to reduce productivity gains. More important, the anti-inflationary fiscal and monetary policies that are invoked, as well as direct controls, tend to erode profit margins, with negative effects on investments. These effects are at least partially concealed by conventional business accounting practices which carry assets and charge depreciation in terms of original costs, and thus overstate profits and rate of return on assets or on net worth computations.

I believe inflation was a significant factor in the productivity slowdown since 1966 and again since early 1973. Further, inadequate expansions of some of our basic industries due to inadequate profit margins has shown up in various materials shortages that developed in 1973, with a further unfavorable impact on productivity even before the oil embargo.

Finally, our market enterprise system has been subjected to increasing guidance, intervention, and control by government. In some cases the social benefits have outweighed the private and social costs, although one doubts how finely the calculus is applied by public officials. There is, for example, a considerable body of opinion that the environmental protection regulations have been pushed beyond the optimum points in some instances. Even if they were not, it is obvious that the significant increases in antipollution outlays in recent years (as well as for occupational health and safety, and other social objectives) have tended to decrease the rate of increase in productivity as measured, since the inputs are included in the productivity ratios while the outputs are generally not counted.

Perhaps the most pernicious longer-term effects on economic and physical productivity come from direct governmental intervention in markets. Our recent almost three years experience with wage and price controls (and to a lesser degree the earlier "guideposts") have demonstrated for all to see the distortions in resource allocation that gradually develop.[5] The earlier interventions in particular industries, particularly agriculture, petroleum, and natural gas, were also a significant element in the recent shortages. It is to be hoped that the lesson has been learned and that relative price and profit changes will be allowed to perform their function of attracting investment resources to expand productive capacity and efficiency.

Finally, a few words about the tax system which obviously affects incentives and sources of funds for investment. There have not been major changes in recent years, but it is my impression that weight is often given to "equity" considerations without due consideration of incentive effects. Yet, over the long run, potential improvements in income of lower-income groups through redistributions are small compared with the potential increases due to productivity improvements with the accompanying rising average income per capita. Two measures adopted in recent years, the investment tax credit and accelerated depreciation allowances, do represent a significant recognition of the importance of adequate incentives.

Quite apart from the structure of the tax system, the total tax burden is also an important consideration. An increasing proportion of income siphoned off by taxes tends to reduce private saving. Unless reduced private saving and investment is offset by increased public investment, and unless the returns to such public investments are equal to private returns, again the productivity impact is negative. These matters deserve more research than they have received.

Total Investment

The most important proximate cause of secular productivity advance is the volume of investment. Here, I define investment broadly to include all outlays that maintain or enhance output- and income-producing capacity. It comprises not only the tangible investments in structures, equipment, inventories, and natural resource development; but also the intangible investments in research and development (R & D), education, training, health, safety, and mobility.

A recent study completed for the National Bureau of Economic Research [6] shows that total gross investment in the United States by all sectors has risen from 35 percent of GNP in 1929 to almost 45 percent in 1969. The tangible investment proportion has remained relatively constant at around 23 percent in good years; it is the intangible investments, designed to enhance the efficiency of the tangible human and nonhuman capital in which it is embodied, that accounts for the relative increase.

Consistent with the relative investment trends, the real stocks of intangible capital have grown relative to the real tangible stocks—by more than 2 percent a year, on average. However, this does not mean that the relative growth of intangible capital accounts for all of the increase in total tangible factor productivity. Over the forty-year period, intangible capital was only about half the size of tangible capital stocks, on average. So, assuming the same marginal value productivity of both types, the relative growth of intangibles would have accounted for about half the productivity increment. Actually, my study suggests that the rate of return on human and intangible investment was somewhat higher than that on tangible investments. Nevertheless, some of the productivity advance must be explained by other factors we come to later.

The time-series analysis is supported by cross-sectional analyses. On an industry basis, average education of workers and research and development spending as a percent of sales are positively correlated with rates of productivity advance. In a recent study, Terleckyj shows that the correlation coefficients are higher when one includes the R & D transferred to an industry through purchases of capital goods and intermediate products from other industries, with the R & D performed directly by the industry. Significantly, the productivity return on the transferred R & D is higher than that on the direct R & D.[1]

Returning to the time series, the relative rate of increase in intangible investment and stock slowed somewhat after 1965. This was due chiefly to the fact that R & D outlays as a percent of GNP leveled out in the mid-1960s at 3.0 percent, and then dropped to 2.6 percent by 1972. The relative decline was due to a cutback in government-financed R & D, although business-financed R & D continued to increase at a decelerated rate. It is my opinion that this was another important factor in the productivity slowdown.

A last comment on investment relates to the importance of reproducible durable capital goods as a carrier of technology. As successive vintages of machinery and equipment (and plants, to a lesser degree) are improved through research and development and engineering, the replacement process alone would result in some improvements in productivity. Beyond that, the rate of net investment and of growth in the real capital stock, reflected in the average age of the reproducibles, would also be expected to have a productivity impact. This hypothesis, first popularized by Robert Solow,[8] has been substantiated by subsequent regression analysis. It provides part of the rationale for the investment tax credit and other measures to stimulate investment.

Noninvestment Forces

We turn now to a final group of noninvestment-related factors behind productivity change. The influence of the business cycle which affects productivity through changes in rates of utilization of capacity has already has been noted. It should also be noted that on an industry basis, the degree of cyclical variability of production is negatively correlated with productivity growth.

The degree of efficiency relative to the potential with a given technology varies widely. The "learning-curve" phenomenon is part of the story. But after productivity levels off under a given technology, actual performance deviates from normal, depending on management efficiency, worker attitudes, and so on. Realization of potential provides an important source of one-time productivity gains, although technological advance is obviously the more important factor in the long run.

The other chief factors will be discussed briefly under four heads: economies of scale; changes in economic efficiency; changes in the average inherent quality of natural and human resources; and, in the case of the business economy, changes in the volume of service rendered by government to business.

Economies of Scale: Since Adam Smith, it has been an article of faith of economists that the growth in the scale of markets brings economies through greater specialization of men, machines, and plants, and the spreading of overhead functions over larger numbers of units. Our studies show that over two or more cycles, trend rates of growth of output and of productivity are positively correlated. A deceleration in economic growth in the 1966-69 period is one factor which explains the productivity slowdown; George Perry estimates it accounts for almost one-third.[9] That it was only part of the explanation, however, is evidenced by the fact that although economic growth decelerated much more between 1953 and 1960 than after 1966, productivity decelerated noticeably less.

On an industry basis, relative changes in output and in productivity also are positively associated as has been demonstrated in my studies and in those of Fabricant before me.[10] This is only partially the scale effect. There is also the reciprocal impact of productivity on output, working through prices. That is, relative productivity changes are negatively correlated with relative price changes, and in the industry sector of the economy relative price changes are negatively correlated with relative changes in sales and output; hence the positive association between productivity and output changes.

Economic Efficiency: Economic efficiency—the allocation of resources and production in conformity with the community's preference—was discussed above in connection with the institutional framework. Here, it should merely be noted that the degree of concentration in American industry has not changed significantly during the twentieth century, while the degree of labor unionization has declined since the mid-1950s. The mobility of resources in response to changes in preferences and other dynamic forces appears to be reasonably good, although improvements could undoubtedly be made in the functioning of both labor and capital markets.

Natural and Human Resources: With regard to the average inherent quality of natural resources, the hoary law of diminishing returns applies to the extractive industries. Yet, as shown in Table 5-1, this tendency has been more than offset by technological advance; productivity has actually increased more than average in agriculture and mining. Further, the consumption of fibers and minerals, including energy materials, per unit of output has declined significantly in this century. Nevertheless, there has been a marked increase in the proportion of imported minerals to total consumption. Efforts to rely relatively more heavily on domestic sources in the future, and even to seek autarchy in energy, may well tend to lower the rates of productivity advance in extractive industry—at least until there are major breakthroughs in new technologies, as in the processing of shale and tar sands, or in nuclear fusion and solar energy.

The inherent quality of human resources I assume does not change over relevant time periods. But changes in the labor-force mix can change average product per man-hour. For example, the great speed-up in the growth in the labor force in the 1965-70 period, due to the bulge of youthful entrants together with some acceleration in female participation, tended to lower productivity as measured, since these groups have a lower value-added per man-hour. George Perry estimates that this factor also accounted for almost one-third of the slowdown after 1965.[11] This factor is obviously self-correcting, and with the marked deceleration in labor force growth beginning in the late 1970s, changes in the age composition will affect productivity positively (although this may be offset by the effect of a slowing economic growth via the scale factor).

Public Services: Finally, business productivity is conditioned by the volume of public services (as distinct from the legal and institutional framework). These include current services such as those of the police, court system, trade promotion, etc., and the services financed by government, namely, the tangible infrastructure such as roads, and education, R & D, public health and safety, mobility, and so forth. These services are difficult to measure. I once calculated roughly that public services to business have grown approximately in proportion to the growth of real business product since 1929.[12] Although difficult to measure. the importance of productivity-enhancing public services to business should not be denigrated.

SOME POLICY SUGGESTIONS

My review of causal forces behind productivity advance implies a whole menu of policy measures that might be proposed to bolster productivity advance in the United States. I shall, however, limit myself to two specific recommendations, one in the area of federal tax policy and one in the area of business accounting practice.

Tax Credit for R & D Investment

At the outset it should be observed that productivity advance as such is not a direct goal of business. Rather, investments designed to achieve unit-cost reduction should, in principle be pushed up to the point at which the rate of return from cost reduction equals the marginal cost of funds.

It should be noted that tangible (reproducible) investments and intangible investments receive different tax treatment. The intangible outlays—R & D, education, training, health and safety—may be charged off as current expense. Plant and equipment, on the other hand, is capitalized with depreciation charged over their economic life. This means that tangible capital costs are somewhat higher with the additional probability of erosion of the buying power of depreciation reserves through inflation. Accelerated depreciation and the investment tax credit, however, have helped to offset these disadvantages in part.

A good case can be made for additional stimulus to R & D investment. Its key role in productivity advance, as the fountainhead of technological progress, has been noted. Further, it is generally recognized that R & D undertaken by particular companies frequently has important external benefits to other companies and the community as a whole. Thus, decisions to finance R & D made by company managements based on estimated internal rates of return, which average well below social rates of return, result in underinvestment in R & D.

A proposal which I endorse is that 20 percent of business R & D expenditures be allowed as an investment tax credit. In case the loss of tax revenue were judged too great by Congress, an alternative proposal is that 40 percent of incremental expenditures on R & D (over and above those of the prior year, or some other base) be allowed as the investment tax credit.[13] These credits should also be allowed for business grants for R & D to educational and other private nonprofit research institutions. A novel feature which I would add is that R & D outlays by manufacturers of producers goods, both capital and intermediate producers goods, be allowed a larger **tax credit**—say 50 percent more—so that the tax credit would be 30 percent on total R & D, or 60 percent on incremental R & D. This more favorable treatment of R & D for producers goods recognizes the dual importance of process and product innovations in those industries. Improvements in the quality of producers goods result in productivity improvements in the purchasing industries. Further, cost-reducing innovations in the manufacture of producers' goods, by reducing prices to purchasers, encourage tangible investments, the carriers of technological progress. The strategic role of producers' goods is substantiated by Terleckyj, as noted earlier, who finds a significantly higher productivity return to the R & D transferred through producers' goods than the return on direct R & D. This may reflect the fact that most smaller firms engage in little direct R & D but are largely dependent on their suppliers for improved technology.

Before leaving the subject of R & D, let me enter a plea for rationalization of the federal government science policy. In particular, I hope that never again shall we witness the sharp drop in federal funding of R & D which occurred between 1969 and 1972, with the attendant sharp increases in unemployment of scientists and engineers. These resources are too precious to waste. A policy of steady growth in R & D support makes much more sense, although changes in the composition of federally funded R & D in accord with government perceptions of needs and relative priorities are to be expected.

Inflation Accounting

My other proposal is that business generally, and the accounting profession in particular, adopt some form of "inflation accounting."[14] This would mean that productive assets—inventories, depreciable plant and equipment, and other property (as well as financial assets)—would be carried on the books at estimated market prices or replacement costs and that inventory and depreciation charges to current expense would also be at current replacement prices.

Some economists are beginning to recommend that all incomes should be adjusted for price-level changes, similar to the "monetary correction" which has become part of the Brazilian economy.[15] But it is not necessary for the business community to wait for some comprehensive form of price-level adjustment before adopting accounting practices which would more truly reflect profits and rates of return than the present accounting practices which are based on original or acquisition costs.

In an inflationary period, present practice overstates profits since depreciation charges understate replacement cost as do the costs of materials consumed in production unless charged from inventory at market price. In 1973, the U.S. Department of Commerce inventory valuation adjustment to corporate profits was $17.3 billion, reducing reported book profits by almost 14 percent.[16] A depreciation valuation adjustment would have resulted in a further major downward adjustment, although estimates are not available.

If assets, particularly depreciable property, were likewise revalued to replacement costs, they would be carried at higher values in times of rising prices, which would further reduce estimated rates of return on assets or on net worth.

What would be the advantage of price-level adjustments in business accounts? Although some managements make rough adjustments for inflation effects on profits, most have some degree of illusion concerning the true situation. Consequently, dividend payouts tend to be higher and retained earnings plus depreciation charges lower than would otherwise be the case; this tends to lower investment.

Profits taxes would also be lower if the tax authorities were prepared to accept the restated earnings estimates—further increasing internal sources of funds. If necessary, enabling legislation should be passed recognizing the legitimacy for tax purposes of inventory and depreciation valuation adjustments to earnings.

Further, federal economic policy-makers would have a more accurate picture of the effects of fiscal, monetary, and other macroeconomic policies on profit margins, and could guide their actions accordingly. For example, the profit squeeze that occurred between 1966 and 1970 was considerably more severe than reported figures indicate, and the subsequent recovery was less robust. Although Washington policy-makers may have been misled by the unadjusted numbers, stock market investors were not, as indicated by the fact that equity prices since 1966 have gone up much less than unadjusted earnings per share.

I realize that there are serious technical problems involved in inflation accounting. One major debate is over the use of specific price indexes versus general price indexes in revaluing fixed assets and depreciation allowances. I would favor the former, but believe the latter is better than nothing. There is not space here to discuss the problems, but the recently formed Financial Standards Accounting Board is considering the matter.[17] I believe that acceptable conventions for adjusting balance sheets and income statements for price changes can be found, and I hope the Board will issue a favorable opinion on the matter within the year.

Improving the Social Climate

Earlier we noted some of the negative social tendencies of the past decade, including increasing antibusiness and antimaterialist sentiment. There is much more that businessmen, and friends of the enterprise system, could do to counter the criticisms. To the extent that the criticisms are valid, business must be ready to correct abuses and to promote reforms to make the system operate even better in the future than in the past. To the extent that criticisms are unjustified, whether due to ignorance or bias, business must be ready to answer them effectively and promote wider understanding of basic economic principles.

I believe the most important challenge in America today for all of us is to strengthen the social fabric so that we can continue to realize the economic and social progress that is possible if we have the consensus and the will. The reason I favor our type of predominantly private enterprise market economy is not just that it is an efficient engine of progress. More important, I agree with the late distinguished professor, Ludwig von Mises, that the diffusion of power which it represents is an effective bulwark of freedom and democracy. Without freedom, material progress is of doubtful value.

NOTES

1. John W. Kendrick, *Postwar Productivity Trends in the United States, 1948-1969* (New York: National Bureau of Economic Research, 1973).
2. Wesley C. Mitchell, *Business Cycles* (Berkeley: University of California Press, 1913).
3. Geoffrey H. Moore, "New Work on Business Cycles," *53rd Annual Report* (New York: National Bureau of Economic Research, 1973), pp. 14-22.
4. Patricia Capdevielle and Arthur Neef, "Productivity and Unit Labor Costs in 12 Industrial Countries," *Monthly Labor Review* (November 1973), pp. 14-19.
5. Jules Backman, "Price Inflation and Wage-Price Controls," in *Business Problems of the Seventies* (New York: New York University Press, 1973), pp. 80-81.
6. John W. Kendrick, *The Formation and Stocks of Total Capital* (typescript in process of review, 1974).
7. Nestor Terleckyj, *Effects of R & D on the Productivity Growth of Industries* (Washington: National Planning Association, forthcoming 1974).
8. Robert Solow, "Technical Progress, Capital Formation and Economic Growth," *American Economic Review, Papers and Proceedings* (May 1962), pp. 76-86.
9. George L. Perry, "Labor Force Structure, Potential Output, and Productivity," *Brookings Papers on Economic Activity*, III, 1971.
10. Solomon Fabricant, *Employment in Manufacturing, 1899-1939* (New York: National Bureau of Economic Research, 1942); John W. Kendrick, *Productivity Trends in the United States* (Princeton: Princeton University Press for National Bureau of Economic Research, 1961).
11. George L. Perry, *op. cit.*
12. Unpublished worksheets on file at the National Bureau of Economic Research.

13. Cf. Michael Boretsky, "U.S. Technolgy: Trends and Policy Issues" George Washington University, typescript, October 1973.
14. Cf. William Blackie, "The Need for Inflation Accounting," *Business Week* (March 30, 1974), p. 16.
15. See, for example, Milton Friedman, "Economic Miracles," *Newsweek* (January 21, 1974), p. 80.
16. *Survey of Current Business* (February, 1974), p. 11.
17. William Blackie, *op. cit.,* p. 16.